FDA-AIChE Workshop on Adopting Continuous Manufacturing 2016

Overcoming the Perceived Obstacles

Bethesda, Maryland, USA
29 February - 2 March 2016

ISBN: 978-1-5108-2180-4

Printed from e-media with permission by:

Curran Associates, Inc.
57 Morehouse Lane
Red Hook, NY 12571

Some format issues inherent in the e-media version may also appear in this print version.

Printed by Curran Associates, Inc. (2016)

AIChE
120 Wall Street, FL 23
New York, NY 10005-4020

Phone: (800) 242-4363
Fax: (203) 775-5177

www.aiche.org

Additional copies of this publication are available from:

Curran Associates, Inc.
57 Morehouse Lane
Red Hook, NY 12571 USA
Phone: 845-758-0400
Fax: 845-758-2633
Email: curran@proceedings.com
Web: www.proceedings.com

TABLE OF CONTENTS

Continuous Manufacturing
Achieving the Vision of Modernizing Pharmaceutical Manufacturing

FDA-AIChE Workshop on Adopting Continuous Manufacturing
February 29 - March 1, 2016

Rapti Madurawe, Ph.D.
Acting Division I Director, Office of Process and Facilities
FDA/CDER/OPQ

U.S. Food and Drug Administration
Protecting and Promoting Public Health
www.fda.gov

Modernizing Pharmaceutical Manufacture

Many of the greatest advances that have been made from the beginning of the world to the present time have been made in the earnest desire to turn the knowledge of the properties of matter to some purpose useful to mankind.

Lord Kelvin

U.S. Food and Drug Administration
Protecting and Promoting Public Health
www.fda.gov

FDA 21st Century Initiative (2004)

PHARMACEUTICAL CGMPS FOR THE 21ST CENTURY — A RISK-BASED APPROACH

FINAL REPORT

September 2004

Objectives:

- ♦ Encourage the early adoption of new technological advances by the pharmaceutical industry
- ♦ Facilitate industry application of modern quality management techniques, including implementation of quality systems approaches
- ♦ Encourage implementation of risk-based approaches
- ♦ Ensure that regulatory review, compliance, and inspection policies are based on state-of-the-art pharmaceutical science
- ♦ Enhance the consistency and coordination of FDA's drug quality regulatory programs

U.S. Food and Drug Administration
Protecting and Promoting Public Health
www.fda.gov

FDA U.S. Food and Drug Administration
Protecting and Promoting Public Health
www.fda.gov

Desired State for Pharmaceutical Quality

A maximally efficient, agile, flexible pharmaceutical manufacturing sector that reliably produces high quality drug products without extensive regulatory oversight

Characteristics of Desired State:

- Manufacturers have extensive knowledge about critical product and process parameters and quality attributes
- Manufacturers control process through quality systems over life cycle, and strive for continuous improvement

Janet Woodcock, CDER/FDA - 2005 and today

FDA U.S. Food and Drug Administration
Protecting and Promoting Public Health
www.fda.gov

Continuous Manufacturing

- Continuous manufacturing is a modern approach to pharmaceutical manufacture
- Has the potential for unprecedented control of product quality
- Offers agility and flexibility of manufacturing to meet patients needs

Continuous manufacturing can be a major step towards the "Desired State"

U.S. Food and Drug Administration
Protecting and Promoting Public Health
www.fda.gov

Continuous manufacturing is
<u>one of many steps</u>
towards quality…

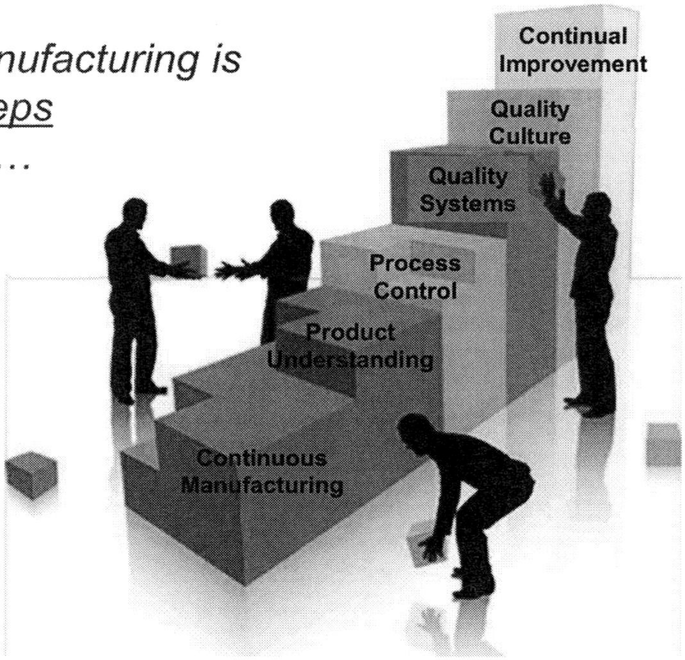

U.S. Food and Drug Administration
Protecting and Promoting Public Health
www.fda.gov

What is Continuous Manufacturing (CM)?

- A manufacturing scheme where the material(s) and product are continuously charged into and discharged from the system, respectively, throughout the duration of the process (Lee, S. L., et. al. (2015))
  - Typically, materials 'flow' between unit operations without product collection at the end of each step
- Most pharmaceuticals made by batch processes
  - Although some unit operations are operated continuously
- Continuous processing of all or multiple unit operations a relatively "novel" concept

U.S. Food and Drug Administration
Protecting and Promoting Public Health
www.fda.gov

Examples of Processes Amenable to Continuous Manufacture

- Drug substance manufacture
  - Flow reactors
  - Separations
  - Simulated moving bed chromatography
  - Crystallization
- Drug product manufacture
  - Blending
  - Granulation
  - Drying
  - Coating
  - Sterilization

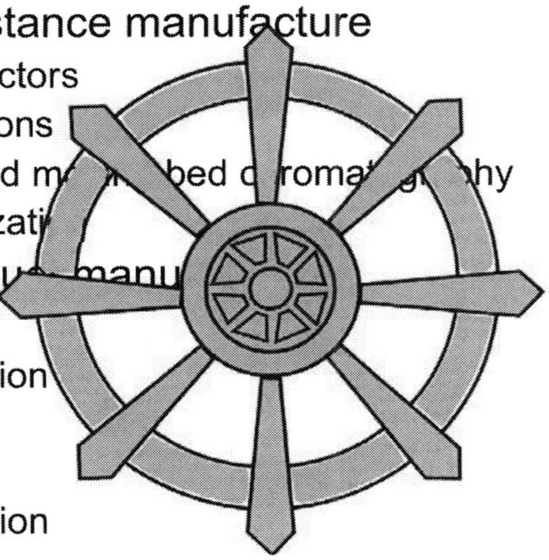

U.S. Food and Drug Administration
Protecting and Promoting Public Health
www.fda.gov

Operational Differences between Batch and Continuous

- Continuous material addition & product collection
  - Flow/throughput rate impacts mixing patterns, process outcomes, sampling frequency, material traceability, etc.
- Connected unit operations
  - Potential for coupled interactions, disturbance propagation
- Potential for process upsets, variability, fouling, etc.,
  - Process monitoring and control
- Rejection of non-conforming material
  - Ability to detect variability and trace material

Control Strategy in Continuous Manufacturing

- Control strategies for continuous manufacturing are usually different than batch
- Controlling process & product quality variation over **time** is a key concern for continuous manufacture
  - Quality variation in **space** could also occur in continuous manufacture, but is generally not as significant a concern as it is for batch processes

U.S. Food and Drug Administration
Protecting and Promoting Public Health
www.fda.gov

Control Opportunities in Continuous Manufacturing

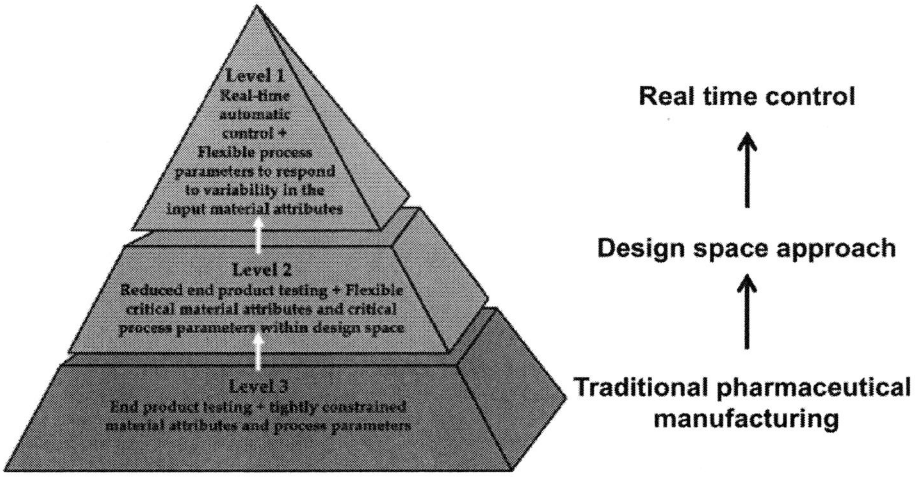

"Understanding Pharmaceutical Quality by Design", L.X. Yu, et. al., *The AAPS Journal*, Vol. 16 (4), 07/14

Advantages of Continuous Manufacturing

- Smaller equipment and facilities
  - Lower capital, operational and environmental costs
  - Increased safety
  - Modular units
- Integrated processing with fewer steps
  - Total processing times shorter; faster product release
  - Simplified scale-up
- More flexible operation
  - Adjust production to meet demand, emergencies, etc.
- Amenable to real time product quality assurance
  - On-line monitoring, control and RTRt approaches

U.S. Food and Drug Administration
Protecting and Promoting Public Health
www.fda.gov

Continuous Manufacturing Development Advantages

- Ability to conduct development studies using commercial equipment
  - Lowered risks in scale-up
- Rapid development screening over many conditions
  - Potential for automated experimentation
  - Potential for less material usage for
  - Ability to build quality into process design
- Ability to use new chemistries or process conditions
  - Highly exothermic reactions, Ultra high/low temperature

U.S. Food and Drug Administration
Protecting and Promoting Public Health
www.fda.gov

Example 1
Continuous Manufacture Benefits

Moffat-Swern oxidation:

- Monitored on-line by Raman spectroscopy
- Batch processes require ultra low temperatures for appreciable yield
- Better yields obtained in continuous flow at higher temperatures

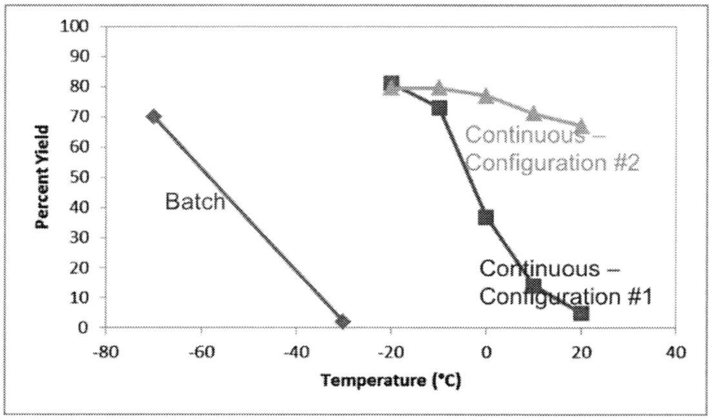

From: "Moffat–Swern Oxidation of Alcohols: Translating a Batch Reaction to a Continuous-Flow Reaction", O. Bleie, M. F. Roberto, T. I. Dearing, C. W. Branham, O. M. Kvalheim and B. J. Marquardt, *J. Flow Chem*, 2015, 5(3), 183–189

U.S. Food and Drug Administration
Protecting and Promoting Public Health
www.fda.gov

Example 2
Continuous Manufacture Benefits

Batch	7 Rooms	2 week production	Make → test quality
CM	2 Rooms	1 day production	Continuous monitoring of quality
Benefits	Increase yield by reducing waste by 33%	Reduce manufacture and testing time by 80%	70% of highest volume products by CM within 8 years

Source: http://www.pharmaceuticalonline.com/doc/janssen-s-fruitful-continuous-manufacturing-collaboration-0001

U.S. Food and Drug Administration
Protecting and Promoting Public Health
www.fda.gov

Drivers for Continuous Manufacturing

- **Cost reduction**
- Reliable and consistent production of quality product
  - Lower product recalls, discarded lots, etc.
- Adjust manufacture to meet demand
- Ability to control process through quality systems over life cycle
- Ability to strive for continuous improvement
- **Benefit to patients**

Going continuous - a company decision

U.S. Food and Drug Administration
Protecting and Promoting Public Health
www.fda.gov

FDA's Facilitation of Continuous Manufacturing

Emerging Manufacturing Technology

Lifecycle Management

Comparability Protocols

Quality Metrics

PV, ICH Q8,9,10 Guidances

Office of Pharmaceutical Quality

Program Alignment Group

" The Desired State "

Quality by Design

Process Analytical Technology

Pharmaceutical Quality Systems

Risk Management

Real Time Release Testing

Knowledge Management

New Inspection Protocol Project

U.S. Food and Drug Administration
Protecting and Promoting Public Health
www.fda.gov

Office of Pharmaceutical Quality (OPQ)

est. January 2015

Centralize quality drug review—creating one quality voice by integrating quality review, quality evaluation, and inspection across the product lifecycle

Consistent quality standards and risk-based approaches

Organization designed to deliver on the promise of the 21st Century Quality Initiatives

U.S. Food and Drug Administration
Protecting and Promoting Public Health
www.fda.gov

Emerging Technology Team (ETT)

- To encourage the adoption of innovative approaches to pharmaceutical manufacturing
  - Cross functional team; representation from all OPQ offices
  - Answer questions on information expected in submissions
  - Identify and help facilitate regulatory review of new technology
  - Serve as lead or co-lead on the quality assessment team
  - Identify and capture resolution to policy issues that may help with future submissions for the same technology
- Refer to ETT Draft Guidance for Industry, 12/2015

  Advancement of Emerging Technology Applications to Modernize the Pharmaceutical Manufacturing Base

U.S. Food and Drug Administration
Protecting and Promoting Public Health
www.fda.gov

FDA Continuous Manufacturing Interactions

- Many public presentations
  - Information on control strategy and regulatory issues
- Meetings with industry and facility visits
- EMA interactions
  - Parallel consultative assessment under the QbD pilot
- FDA sponsored research and academic collaborations
- Collaboration with BARDA to further the use of CM
  - Biomedical Advanced Defense and Research Authority
- Staff training, communities of interest, etc.

U.S. Food and Drug Administration
Protecting and Promoting Public Health
www.fda.gov

Requesting a Meeting

- **IND/NDA**: Can request a "Type C meeting "*Formal Meetings Between the FDA and Sponsors or Applicant*"
  - Can request to participate in the ETT program meeting http://www.fda.gov/downloads/Drugs/Guidances/ucm153222.pdf
- **ANDA:** Can submit a pre-ANDA meeting request
  - Specify the meeting request as a "Pre-ANDA meeting – request to participate in the ETT program"
  - Send to CDER-ETT@fda.hhs.gov
  - Submit at least three months prior to the application date
  - FDA will respond in writing within 60 days
- Include questions and sufficient background information in meeting package

U.S. Food and Drug Administration
Protecting and Promoting Public Health
www.fda.gov

Current Status of CM

- Perception
  - Regulatory agencies will ask too many questions
  - Not approve CM applications the first review cycle
- Reality
  - One approved NDA, 1st cycle – priority review, EMA QbD pilot
  - Multiple pre-submission meetings (including with EMA) to share information/expectations and reach common ground
  - Questions help clarify unclear areas and facilitate approval
- Industry interests: Multiple companies doing CM
  - Both new molecular entities and existing products
  - For drug substance, drug product, and both integrated
  - Fully and partially continuous process trains
  - Drugs for different routes of administration

U.S. Food and Drug Administration
Protecting and Promoting Public Health
www.fda.gov

U.S. Food and Drug Administration
Protecting and Promoting Public Health
www.fda.gov

Concluding Remarks

- The FDA strongly supports advancement of pharmaceutical manufacturing technologies, including continuous manufacturing
- Continuous pharmaceutical manufacturing is expected to benefit both patients and industry
- FDA recommends early and frequent discussion of advanced manufacturing and analytical approaches with the Agency prior to submission
- FDA is fully staffed, capable of and ready to review continuous manufacturing applications

U.S. Food and Drug Administration
Protecting and Promoting Public Health
www.fda.gov

Thank you!

Please send questions or comments to
CDER-OPQ-Inquiries@fda.hhs.gov

Disclosures

- The speaker is solely responsible for the content of this presentation unless otherwise acknowledged
- The views presented here do not necessarily represent the views of GSK

Acknowledgements

The following people contributed to the ideas expressed in this presentation

- Andy Scott, GSK
- Mike James, GSK
- Hannah Robinson, GSK
- Lindsay Wylie, GSK
- Matt Popkin, GSK
- Gurjit Bajwa, GSK

Continuous Technology Adoption

Continuous Manufacturing is gaining momentum...

API Continuous Manufacturing

OSD Continuous Manufacturing

PAT and Real-time Release

(most companies use PAT but these ones stand-out in terms of application to continuous)

Disclaimer: identification of company activity in continuous manufacturing has not been verified and is based on press releases and publications

Purpose of a File

Patient centric view...when doing something different it is good to get back to basics

New Drug Application (NDA)

The New Drug Application (NDA) is the vehicle through which drug sponsors formally propose that the FDA approve a new medicine in the U.S. The data gathered during the animal studies and human clinical trials of an Investigational New Drug (IND) become part of the NDA.

The goals of the NDA are to provide enough information to permit FDA reviewer to reach the following key decisions:

- Whether the drug is safe and effective in its proposed use (as described in the label) and whether the benefits of the drug outweigh the risks.
- Whether the drug's proposed labelling (package insert) is appropriate, and what it should contain.
- <u>Whether the methods used in manufacturing the drug and the controls used to maintain the drug's quality are adequate to preserve the drug's identity, strength, quality, and purity.</u>

http://www.fda.gov/Drugs/DevelopmentApprovalProcess/HowDrugsareDevelopedandApproved/ApprovalApplications/NewDrugApplicationNDA/

Tiered Approach to Regulatory Engagement

Ensure consistency of message

Industry

- MIT Conference 1
- 2015 DIA/AAPS CMC Workshop
- IFPAC 2014
- DIA 2015 51st Annual Meeting
- AIChE FDA -PD2M conference

Platform

- ISPE China Annual Spring Conference
- 50th AAPS Arden Conference
- QbD meeting at PMDA
- QbD Case Study Group meeting at PMDA
- Pharmaceutical Quality Forum
- INTERPHEX Japan
- APS Pharm Sci Conference, UK
- Invitations to regulators to GEMBA platforms

Project

- Initial FDA meeting
- FDA-ETT
- Direct regulatory engagement related to filing strategy

– Acknowledgements: Andy Scott and Mike James, GSK

Topics for Platform / Project Regulatory Engagement

Different levels of feedback from simple to more complex

Spectrum of regulatory feedback

Definitions → **Clarifications** → **Fundamental**

Definitions

- Process Summary
- Equipment for Continuous Process
- Batch Definition
- Yield
- Raw Materials and Intermediates
- Traceability
- Specifications
- Comparability
- Process Validation/Verification
- Lifecycle Management

Clarifications

- Equipment Design
- Control System Interface
- Characterisation
- R&D Development Campaign
- Residence Times and Dispersion
- State of Control
- Product Collection or Rejection
- Process Monitoring (PAT) and Sampling
- Process Dynamics

Fundamental

- Deviation management
- Control Strategy
- State of Control

Equipment Characterisation

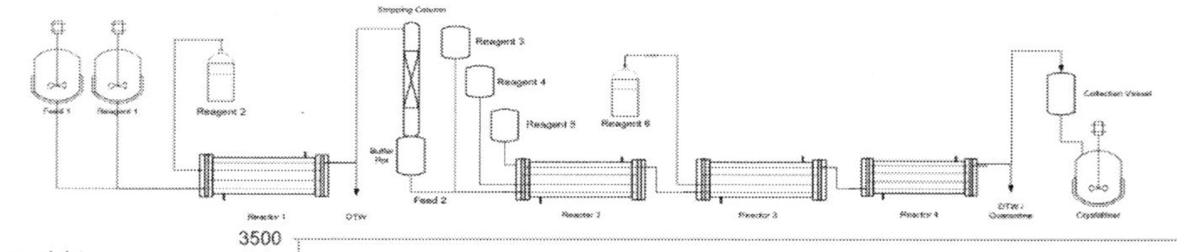

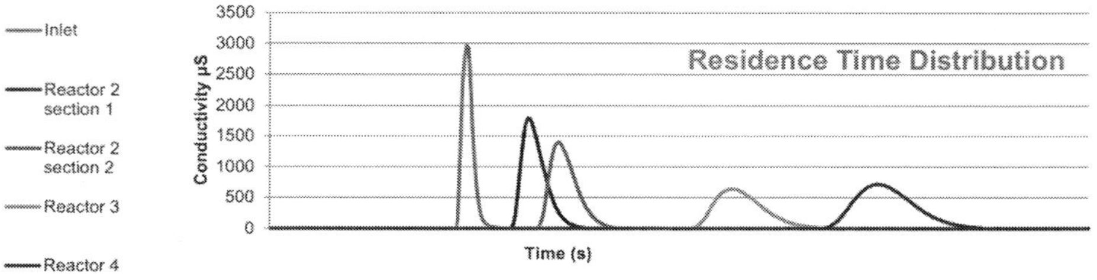

- Mixing intensity and Re
- Reactor volumes and surface area
- Heat transfer capability
- Residence time distribution

Acknowledgement: Hannah Robinson, GSK

Traceability in a Synthesis Flow Process

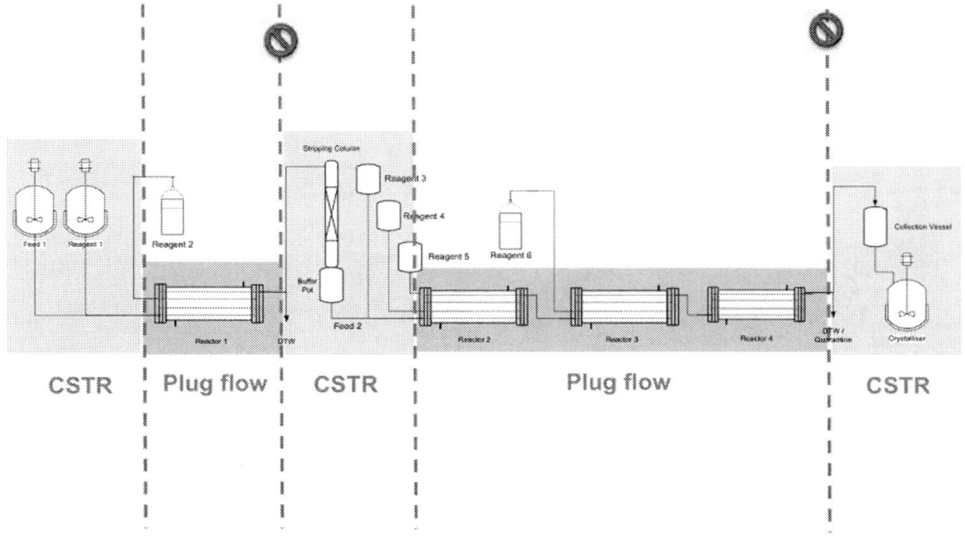

Acknowledgement: Hannah Robinson, GSK

Control phases

How will disturbances be handled?

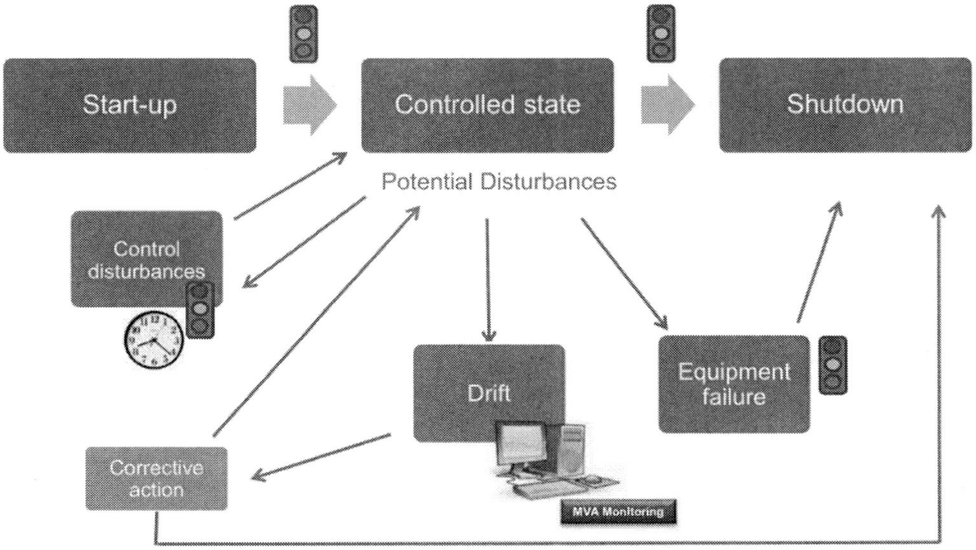

Acknowledgement: Hannah Robinson, GSK

Some thoughts on Control Strategy

Deviations, process dynamics, control loops, consistent output

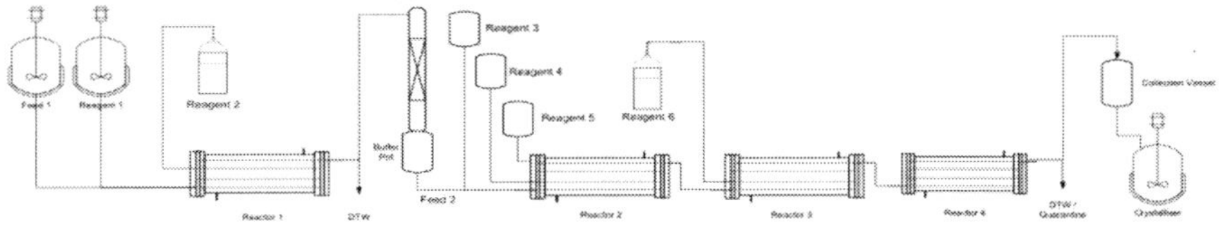

Deviations

- In both batch and continuous deviations need to be investigated and managed – challenge in continuous is that this will need to happen in real-time (at least for process deviations)

Process Dynamics and Control Loops

- "If the process control system is having to actively adjust process parameters to remain within design space or state of control, isn't that a sign of a non-robust process?"
- Reality is that pharma continuous processes will be better understood and controlled – mainly as disturbances related to changes in raw materials will be managed out through the control strategy and equipment configuration; e.g. loss in weight feeders will adjust rpm to deliver desired target throughput rate

Consistent Output

- Precision and accuracy
- Process parameter trends and MVA modelling can demonstrate "precision" – consistent output (whatever that may be)
- PAT enabled as a key component of the control strategy leading to RTRT

Moving from Engagement to Filing

Communication and knowledge management

- Important to link Industry and Platform level engagement to Project level engagement and filing.
- It is at the Project filing level that the fundamental debates need to crystallise unambiguously
- Most issues are around the Product Control Strategy, the role of end-product testing and practicalities of product release

"Tested and complies"

- Batch paradigm
- Assumes uniformity of product which relies on fully blended nature of a batch
- Straight forward to file
- Practically for a QP it is unambiguous and straight forward to release

?

"If tested will comply"

- Continuous paradigm
- Need to demonstrate uniformity of product and complies with specification
- Less clear how to file
- Need to clarify regulatory expectations of both assessment/review and inspection

Conclusions

Continuous Manufacturing is gaining momentum...

- Best practise is to create a framework for regulatory engagement that covers Industry, Platform and Project levels
- Need to actively manage the link between regulatory engagement and filing to drive consistency; early and continued dialogue with regulators is important
- The most important thing to crystallise is approach to Product Control Strategy and the shift in paradigm from "Tested and complies" to "If tested will comply"
- There shouldn't be any regulatory hurdles to implementation of CM; only points for clarification
- From industry perspective it is important to harmonisation of regulatory, filing and inspection approach

Design and Implementation of a Small Footprint Continuous API facility for Portfolio Commercialization

Eoin McManus
Eli Lilly and Company

March 2016

Agenda

1. Drivers for Continuous API
2. Phase 1: Platform Technologies & Reactions
3. Phase 2: Proof of concept with Manufacturing
4. Phase 3: Proof of concept Small volume continuous
5. Phase 4: Construction of Novel Facility
6. Overall Learnings

What are the Drivers for Applying Continuous Processing?

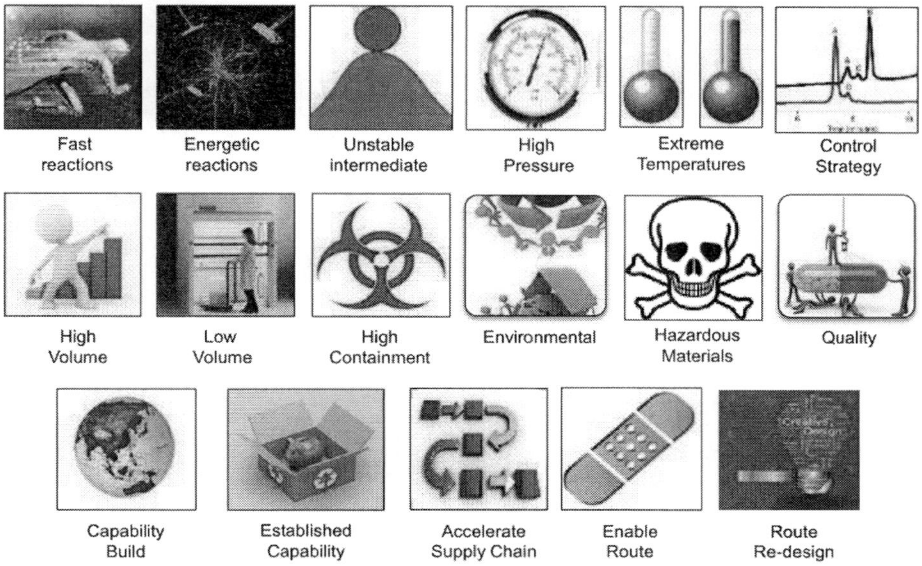

3

Realized Benefits of Continuous at Lilly

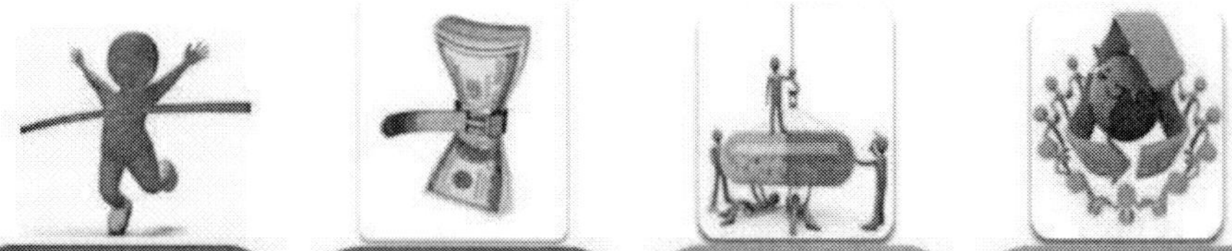

Improved Speed
- -Automated DOEs
- -Eliminate unit ops
- -Streamline tech transfer
- -Rapid prototyping

Lower Cost
- -Better synthetic routes
- -New capabilities & added capacity
- -Deferred capital investment
- -Increased throughput

Improved Quality
- -Real-time analytical
- -Steady State
- -Equipment is engineered to process
- -Less material at risk

Improved HSE
- -Reduced waste
- -Containment
- -Smaller volumes
- -Lower reagent amounts
- -Neat reactions

Timeline for API continuous at Lilly

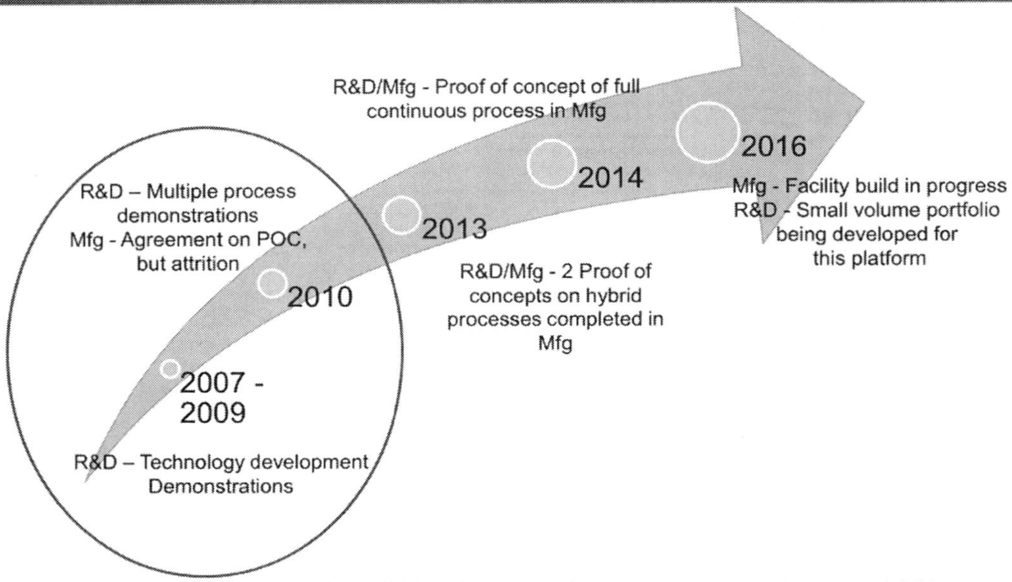

Phase 1 : Platform development in R&D

Platform Technologies for Continuous Reactions

80L PFR

<u>Plug Flow Reactors (PFRs)</u>

1. High or low temperature and pressures with all liquid reagents and product
2. High pressure hydrogenation, carbonylation, aerobic oxidation with liquid + gas reagents
3. Azide formation, thermal cyclization and deprotections, cryogenic reactions
4. Packed bed reactions

200L Pipes in Series
Demo model only
(Production Scale)

6L CSTR

<u>Continuous Stirred Tank Reactors (CSTRs)</u>

1. Low or high temperature and pressure with solids in flow
2. Organometallic reactions including Grignards
3. Biphasic coupling reactions or sequestered catalyst

Platforms for Continuous Workup/Isolation

Timeline for API continuous at Lilly

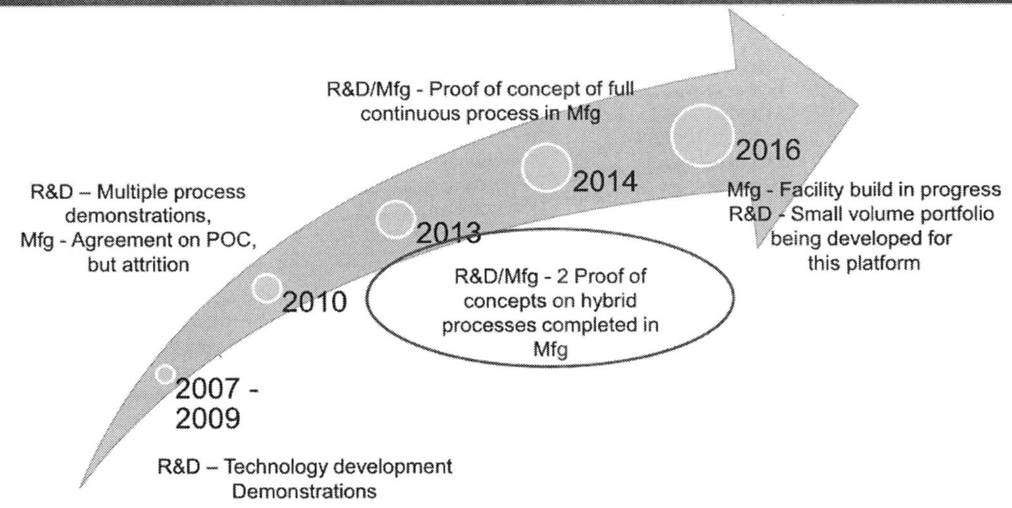

Phase 2: Proof of concept within manufacturing (Hybrid process)

Proofs of concept in Lilly Manufacturing

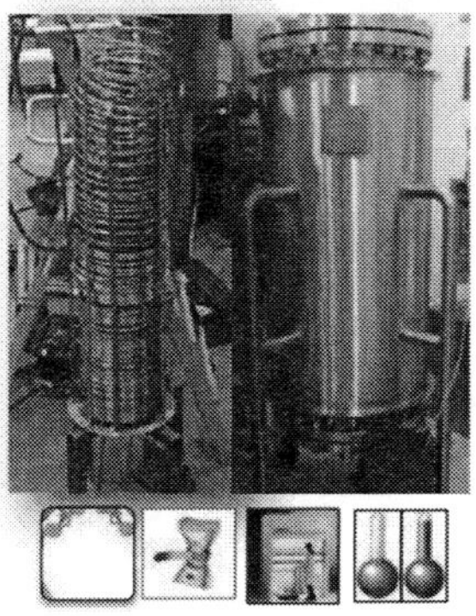

Project A
API cost reduction of >30%
Robust Pd removal
Green chemistry reduces waste

Registration Stability
120 kgs

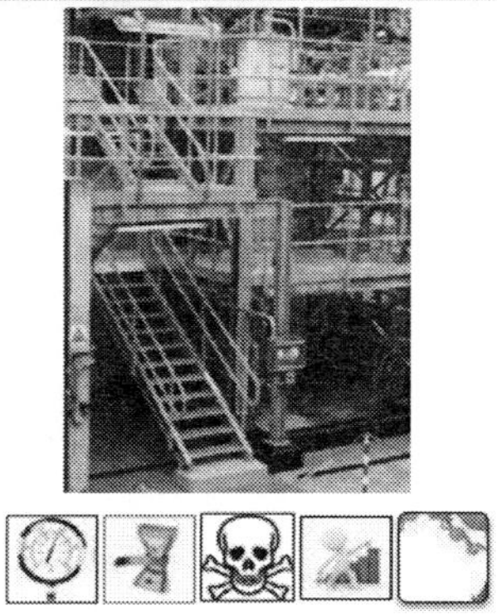

Project B
Eliminated \$20M spend on H_2 bunker
Safer alternative to batch
Green chemistry

Registration Stability
1800 kgs

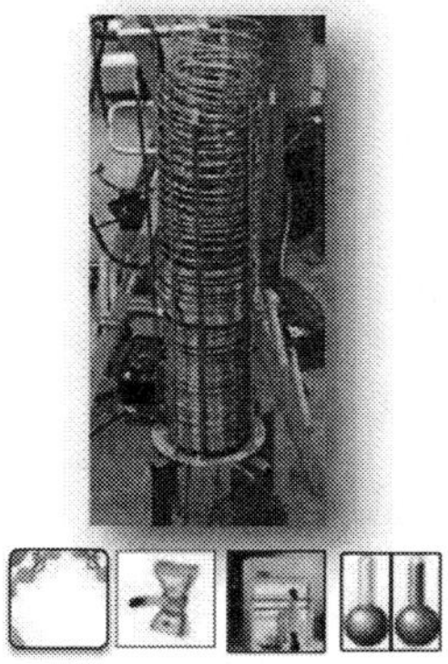

Project C
Development time reduced
Tech transfer req'd half the time
Green chemistry reduces waste

Pre-registration stability
150 kgs

Case Study: Continuous Unit Operations – Thermal PFR and adsorption

The filling of the feed tanks with a new T8 batch will allow batch definition through the continuous process

Palladium Adsorption

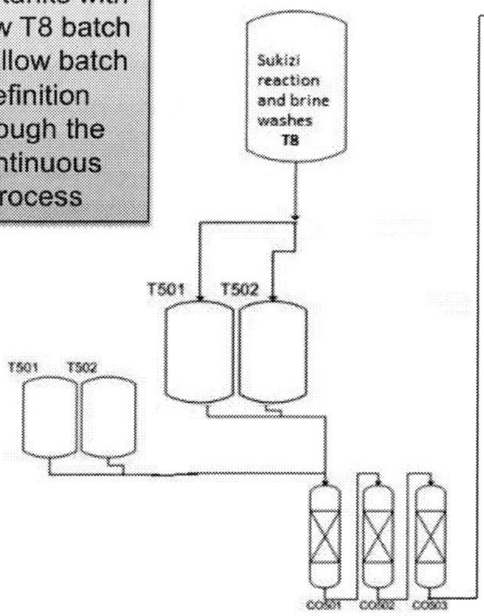

Packed bed column for Pd removal

De-BOC reaction

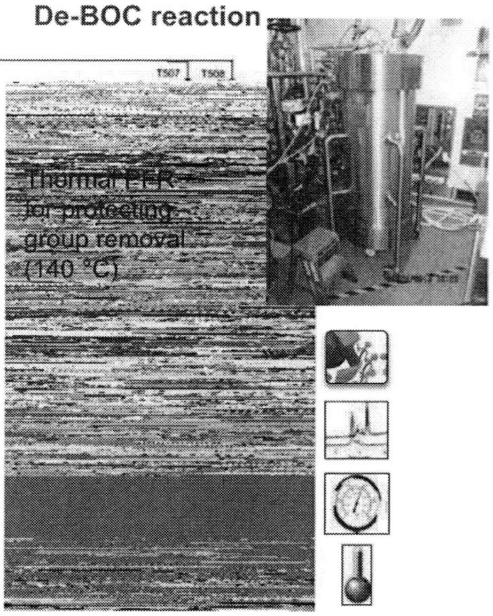

10

Realized Cost Savings

Process Element	Quality and Technical Rationale	Benefits for Primary Stability Campaign	Commercial Processing Benefits
Column Pd removal process	•Minimizes use of expensive resin to remove Pd catalyst •Robustness of impurity control	↓ API cost (6%)*	↓ API cost (10%)*
Plug flow reactor	•Readily achieves desired high temperature •Enables elimination of tech to final crystallization step	↓ cycle time: 2 weeks ↑ Overall yield 10%	↓API Cost (15%)*

*Cost reductions are based on R&D costing models and are relative to the all batch process used in pivotal campaign. R&D's costing models and are used to help guide process development.

11

Case Study: Safety and High Volume Drivers for a Reductive Amination

- *High projected peak volume (>100,000 kg/year)*
- *High Pressure Hydrogenation has a superior safety profile* vs. batch alternatives
  - Runs as a *low risk operation*
  - All hydrogen is kept outside of the building
  - 380 L pipes-in-series reactor – Lilly design
- *It is a green alternative to the previous approach which used stoichiometric sodium triacetoxyborohydride (STAB).*
  - Avoids the use of 1.2 million kg of STAB over the lifetime of the product
  - *Catalysis* – Just 0.001 eq of catalyst used with option to recover and recycle the metal.
- Significant cost savings in $/kg API

Material Comparison in 2 metric ton Campaign (Kinsale, Ireland 2013): 983 kg STAB vs 1.1 kg Ir catalyst

12

Pipes in Series Reactor for Hydrogenations

380 L Reactor – Kinsale IE2 facility

13

Safety & Capital Cost Benefits

1. Safety
    - >99% liquid filled reactor with segmented flow of hydrogen vs. batch where 30% of headspace is hydrogen
    - 50X less hydrogen in the system at any point in time

2. Capital Avoidance
    - Batch: 2500L autoclave rated to 1000 psig and bunker = $30MM
    - Continuous: 850L plug flow pipe system = $2MM
    - $28MM potential CapEx savings

3. Other factors can heighten cost risk
    - Uncertainty around approval of the drug
    - Uncertainty around peak demand of API
    - Uncertainty around long term use beyond the lifecycle of the product

14

Timeline for API continuous at Lilly

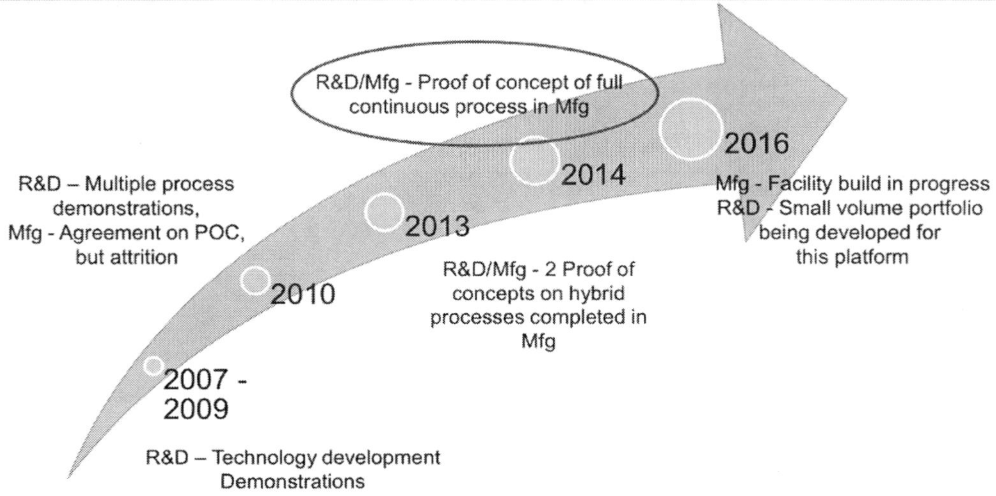

Phase 3: Proof of concept Small Volume Continuous

Small Volume Continuous Demonstration at GMP Manufacturing Facility

- Asset was low volume oncolytic perfect for SVC concept
- Continuous process reduced cost by 57% and waste by 62% ; also eliminated 2 crystallizations and reduced handling of potent compounds

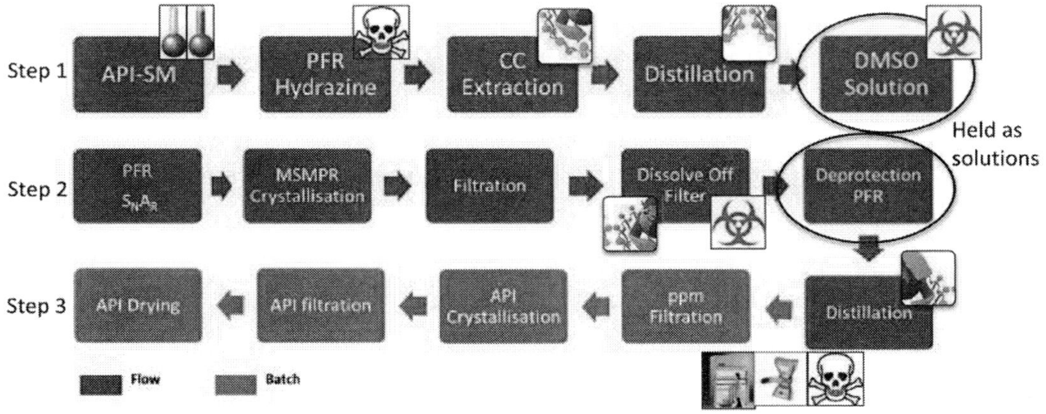

cGMP Campaign: Step 1

Reaction + Monitoring

Continuous Extraction

Automated Distillation

- This entire step occupied 3 standard size walk-in fume hoods.
- Continuous unit operations significantly reduced hydrazine concentration in reaction and allowed its removal to ppm levels.
- On-line analytical verified that operations were in a state of control.
- Ran for over 200 hours at 3.3 kg/day.

17

cGMP Campaign: Step 2

Reaction + Monitoring

Continuous Crystallization

Filtration Dissolution

Reaction + Monitoring + Pumps

- This entire step occupied 3 standard size walk-in fume hoods.
- On-line HPLC units were run simultaneously to monitor the purity of each reaction.
- Ran for over 200 hours at 3.3 kg/day.

cGMP Campaign: Step 3

Distillation

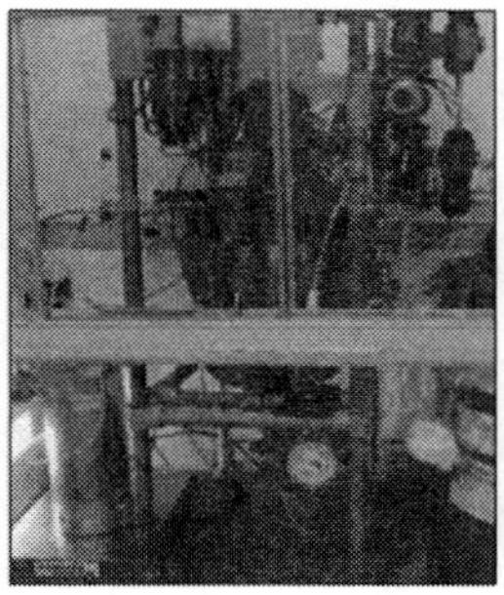

Holding Tank

Batch Crystallization

Drying

- Distillation conducted semi-continuous feeding into holding tank
- Final crystallization with filtration and drying of low OEL compound performed batch
- 5 GMP batches isolated

19

Process Analytical Technology

- Used on-line uPLC (Waters PATrol) after each reaction, this was crucial to monitoring the health of the process.
- Refractive index utilized to measure axial dispersion and good indication of steady-state

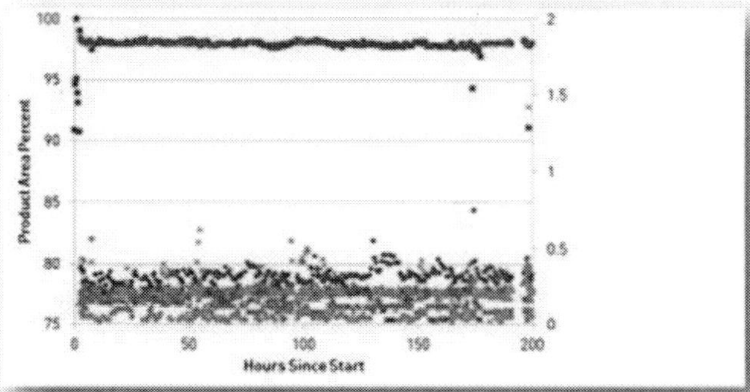

uPLC response corresponded to temperature for reactor.

Monitored to ensure impurities remained at acceptable levels.

Process Monitoring

- Data historian (PI) for monitoring
- Scripts were running to send alerts to personnel when something should be checked.

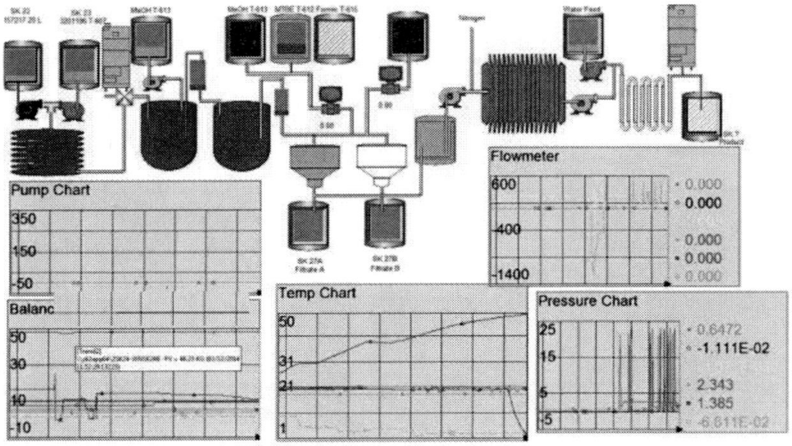

Timeline for API continuous at Lilly

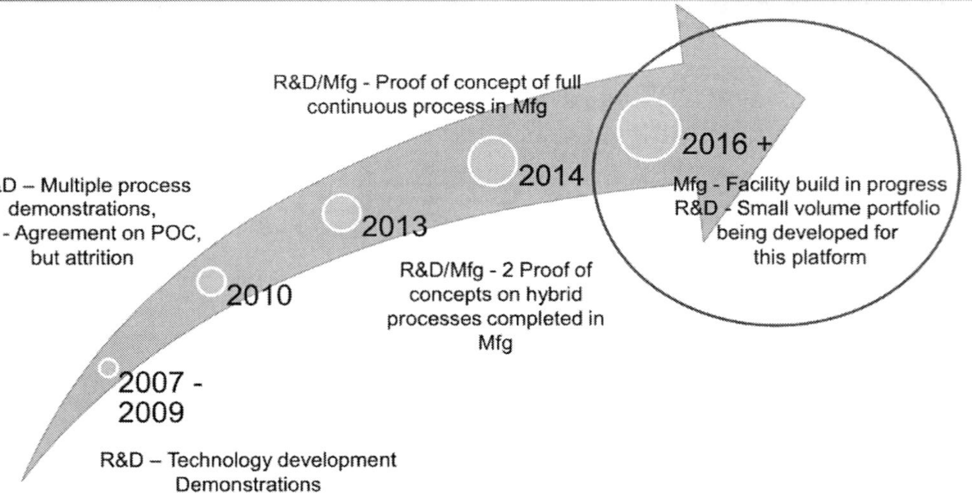

Phase 4: Small Volume Continuous platform implementation

Why Small Volume Continuous ?

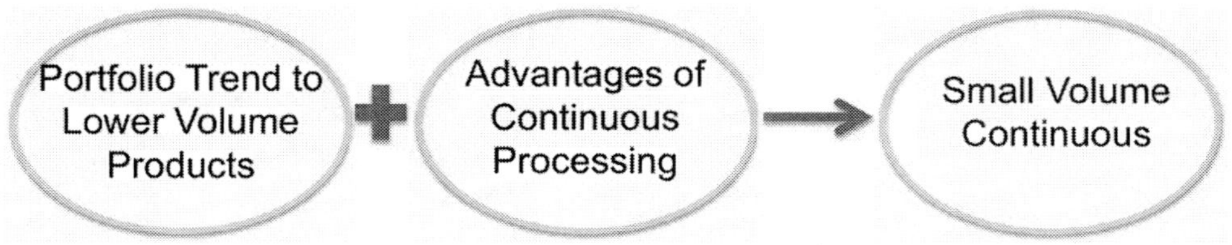

- Small Volume Continuous is the concept of fully continuous processes operating at rates of 3-10 kg/day to deliver material quantities less than 1500 kgs/yr
- >75% of Lillys post-FHD portfolio has a projected API volume of less than 1500 Kgs/yr
- Fully continuous processes at these scales allows for the following benefits:
  - Enhanced plant productivity
  - Reduction in capital investment
  - Fast start/stop decisions
  - Wider p/T operating space afforded with continuous platforms
  - Risk reduction since process is developed/demonstrated at commercial scale

Material Throughput Advantage of Continuous over Traditional Batch Processes

As an example, consider producing 400 kg of API via a batch process:

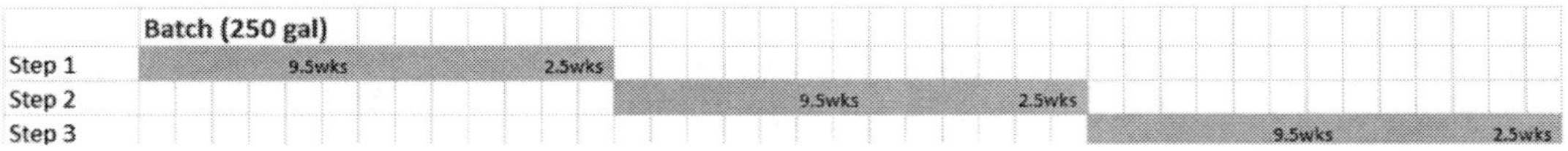

Batch throughput is 6 kg/day; average of Kinsale campaigns

Using a fully continuous process where all the steps are running simultaneously:

IE2 SVC Facility – Operational in 2017

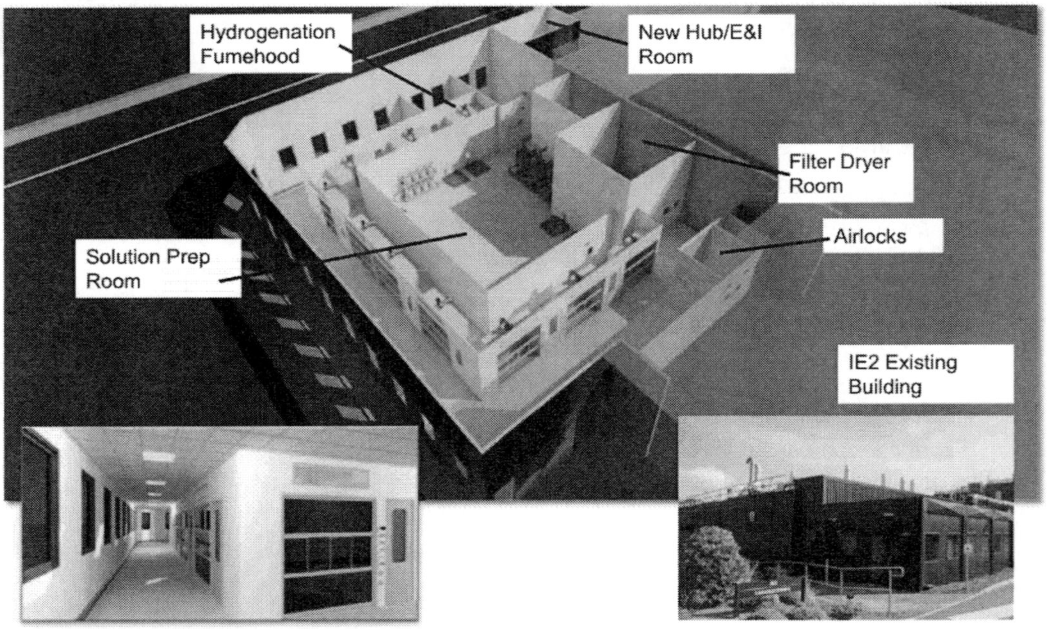

Reshaping the Irish Landscape

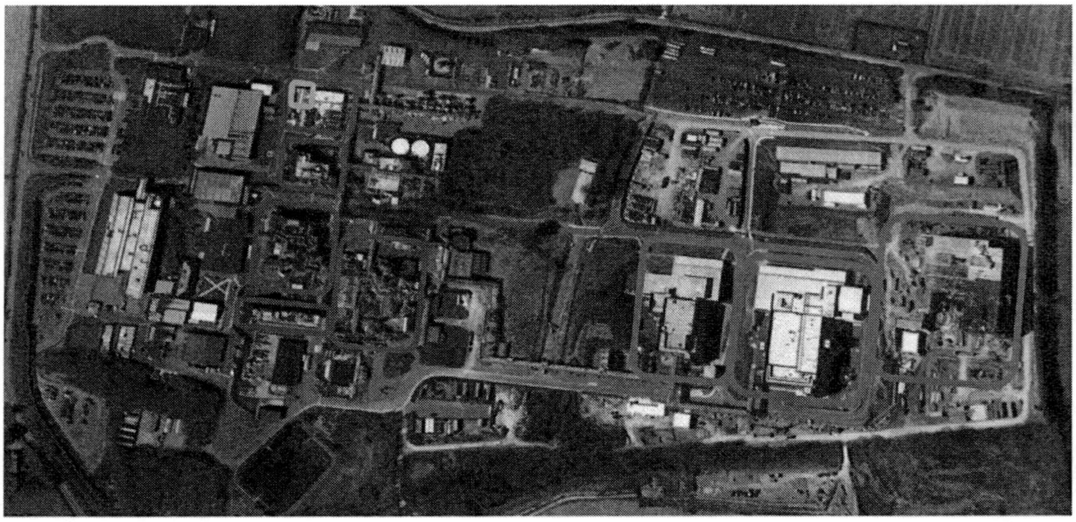

Program for Small Volume Continuous Platform

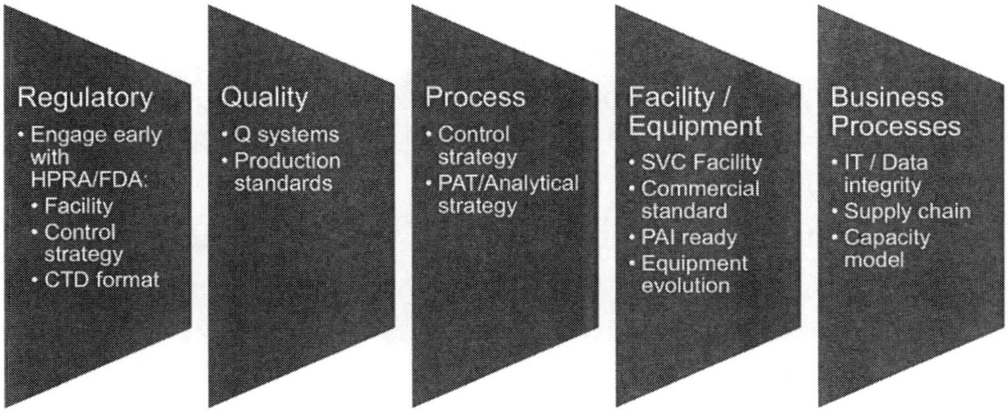

Significant program to enable Small Volume Continuous platform
Priority initiative in parallel to Facility delivery 2015-2017

27

API Accomplishments to Date

Lilly manufacturing:
- 2 registration stability campaigns in 2013
- 2 clinical trial campaigns including 1 SVC demonstration in 2014
- 1 validation campaign in 2015

External manufacturing:
- 2 clinical trial campaigns in 2013
- 2 validation campaigns in 2015

Regulatory activity:
- 1 Type C meeting with FDA in 2013
- Several End of Phase 2 (EOP2) with FDA meetings since 2013
- 1 NDA submission with FDA in 2016 and 1 further submission planned (for processes with continuous unit ops)

Overall Learnings

- Continuous processing has many benefits, but business case has evolved over time
- Each major internal partner (R&D, Manufacturing, Quality, Regulatory etc.) had a different driver
- Leadership and trust needed
- Close partnership between R&D and Manufacturing
- Proof of Concepts invaluable to gain alignment
  - Each also had individual business case
- Portfolio attrition delayed manufacturing POCs
  - Selected 'high probability of next campaign'
- Regulatory feedback lagged implementation
- Third party network is essential

29

Acknowledgements

Derek Berglund

Scott May

Marty Johnson

Jen Groh

Tim White

Paul Collins

Bret Huff

Ciara Hood

Kinsale Ireland Operations Staff

30

Report-out on Topic 1:

Perceived Challenges with FDA/EU/Other Regulatory Review Questions, as well as PAIs

Version 1:

As presented on Day 2 of Workshop

Ground Rules for Discussion

- Intended as a technical discussion
- Intended to facilitate sharing of experience
- Not a Q and A for the Agency
- Questions regarding specific applications or requests for a meeting can be sent to: CDER-OPQ-Inquiries@FDA.HHS.GOV

Communication and openness

- Internal discussions within companies facilitated internal alignment around a lot of the CM related aspects (e.g., batch definition). Multiple companies shared that visits from regulators to sites for CM equipment tours and associated discussions were very fruitful.
- The agency is advocating early and frequent discussions. This will help both the sponsors as well as the regulators to advance the learning and ensure that both are learning at the same time.
- While there may be a number of questions for the initial filing, a lot of the learning will be recycled towards platform understanding and the general expectations from the audience that the number of questions will be lower for subsequent filings as the platform and its understanding becomes more mature.

Shared learning

- Educational visits to operational CM facilities considered very beneficial and worthwhile
  - To educate internal colleagues
  - To educate regulatory agencies
  - To demystify what can otherwise seems like an abstract, potentially complex and challenging mode of manufacture
- With CM, (*a factor that reduces risk is*) that the same equipment set is used for development through commercialization
- CM offers opportunity for increased process knowledge - could permit easier / faster post approval changes in the future
  - Current filing is potentially too static
- With CM, it is expected to have to deal with much larger data sets than conventional batch processes. Clear understanding and expectation is needed to deal with data traceability, data integrity, and ability to address conflicting data from multiple data streams or to deal with aberrant data points.

Shared learning, continued

- Opportunities to share generic learnings from CM experience
  - Pre-competitively
  - Possibly via a wiki or database with some curation
  - Considered potentially sustainable compared to large consortium effort
  - Possible vehicles via IQ or similar
  - A repository of information on what models work well for what operations also considered valuable
- Questions on filings are part of this shared learning
  - Evidence that those questions add value
- The value vs cost/risk of redundant controls in CM:
  - is it possible to utilize all data streams initially to gain more knowledge and then transition to fewer controls long term?
  - A proposal that it may be easier to drop some controls if the sponsor can prove that the associated failure modes are already covered by other controls.
  - This may be easier to implement this way rather than relying only on historical data to justify fewer controls.
  - Different views were also shared to show the potential advantage for redundant controls (e.g., ability to run if one layer is not functioning while other redundant layers are).
  - While it may not be a regulatory expectation to have multiple, redundant controls with CM, it is up to the sponsor to determine the need for redundant controls to assure robustness of the process and product quality.

Do we need guidance? Are we ready for guidance?

- Is shared experience sufficient to support drafting of guidance?
  - Mostly "no" but a few "yes"
  - Have we seen all the innovations that deserve fair consideration?
- Interest in seeing more case studies in the literature
- More forums allowing collaboration and sharing
  - Having these forums is very helpful
  - White paper could provide ~summary of these conferences for those who don't attend

Overcoming perceived barriers

- Perceived barriers– what agency wants vs what industry thinks agency wants: How to achieve predictability?
- Probably the biggest perceived risk was global regulatory acceptability.
  - However, the participants in the discussion agreed that it was difficult to identify any real life examples where an international regulatory body might have impeded implementation of an innovative technology. Nonetheless, the participants agreed on the value of sharing data as well as proceedings from a conference like this one with the global audience, as well as conducting tours/visits to further lessen this perceived risk.
- Some of the perceived barriers were whether CM was approvable. Now that a CM process has already been approved, that perceived risk has been lowered.
  - Some of the perceived barriers within individual companies are around the fear that CM will not be approved or that it wont be approved within the specific project timeline.
- Perception that regulators have narrow experience in CM applications
- Industry: concerned that CM will follow same pattern as QbD
  - Initially, shared learning and many questions
    - Some questions were not possible to answer, not relevant; # of q's is not always the problem
    - The quality of questions is improving, more pertinent
    - Not always clear on which questions driven by technical curiosity, which are roadblocks to approval
  - Scientific Understanding a given, but regulatory flexibility elusive
  - Incorporated lessons learned – to improve or adjust presentation of data
  - Industry and regulators to build on this to improve experience for CM

Overcoming perceived barriers

- Fear that if provide wrong data will be locked in to untenable constraints
- Regulatory harmonization: willingness of regulatory authorities outside ICH to support CM processes
  - Improve communication?
- Knowledge management: multiple dimensions
  - Tons of data: how to manage?
- Sponsorship and support within companies?
  - Forums exist where high level people attend
    - There is a range of approach – from new idea welcome to "show me robust roi etc"
  - How to balance risk fear opportunity etc?
- How much change is needed for internal quality systems to adapt to CM? An example was shared where a company reviewed all QMS and only a handful needed to be revised.

Overcoming perceived barriers

- Practitioners indicated that initial skepticism had been overcome by the practical experience of seeing CM really working
  - No longer phased by potential complexity or data volume
  - Seeing opportunities to take advantage of several drives in multiple processes
- Company culture issues
  - Considered important to have chemical engineers in the right roles in order to facilitate uptake of CM

Challenges for submission and validation

- Have deep technical knowledge about your process and controls
  - this knowledge partly counteracts risk of bringing on new tech
- How much detail to include?
- Concerns on peoples' minds
  - How to handle outliers in real time __?
  - Differentiating between robustness of system and controls needed to keep system stable
- Interaction is good
  - Conversation with agency addressed this
  - Feedback loop, lessons learned helps
  - Where to put post approval changes?
  - Agency is not too picky about what goes in which sections
    - do orgs have/will produce cheat sheets for what goes in what sections? Useful?
- Validating CM processes
  - Aligning CM processes with validation guidances in US and EU
  - Continuous validation and verification – what does this mean?
  - What constitutes validation? Are there different ways/strategies/approaches?
    - Acknowledging Risks of scale up/scale out
- Need to learn by doing.

Possible actions

- More visits to operating CM facilities
- More industry publications of learnings, pre competitive
- More workshops and forums of this type
- Possible white paper from regulators
- Possible wiki or other vehicle to share
  - Potential to be global, could assist challenges in other regulatory regions

- Other dissemination opportunities at forthcoming meetings in 2016
- Multiple, separate efforts are ongoing to capture approaches for CM, including this workshop, CMAC, the upcoming ISPE conference, IQ, as well as other consortia.
  - There will be a great value in aligning different buckets of work that are being addressed separately in these different forums so that topics are distributed and the various efforts complement each other rather than competing on the same subjects.
- (Direct follow on workshop
  - in Ireland Feb 2017 led by SSPC: manufacturing focus)

Notes of Topic 1 report-out discussion

- Extra items to add:
  - Modeling
    - Should be included as a section in the wiki
  - Validation
    - how to determine the right running time (vs 6 month run time in practice)
  - iQ and SSPC willing to work on wiki
    - Some concerns about ensuring information quality
    - Valuable to also share questions received on filings
  - Important for the resource to be open to all
    - Industry, regulators, academics

Session II: Top Challenges Faced in Industry based Focus Groups

AIChE FDA Workshop on Adopting Continuous Manufacturing

Topic 2: Applicability of Continuous Manufacuring to a Diverse Pharmaceutical Portfolio

1 Draft 22/10/2015

What is a diverse pharmaceutical portfolio?

Drug Substance

- Therapeutic area
- Phase of Development
- Production volume
- Containment & potency
- Project timelines and acceleration (breakthrough status)
- Molecular complexity
- Variety of reagents / transformations
- Complex control strategy
- Rate of reaction
- Solid state form
- Physicochemical properties
- Bulk/powder properties

Drug Product

- Therapeutic area
- Phase of Development
- Production volume
- Containment & potency
- Project timelines and acceleration (breakthrough status)
- Route of administration
- Dosage forms / release profiles
- Unit dose strengths
- Manufacturing platform
- Combination products

What are the benefits & barriers of implementation of continuous processing?

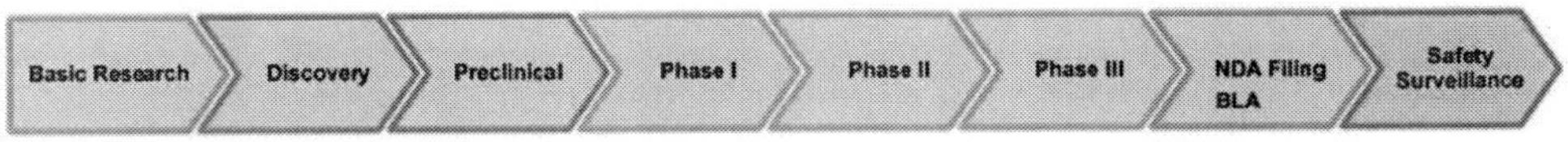

Benefits

	DS	DP
Development Efficiency		
Design for continuous early	X	X
Expanded chemistry tool box	X	
Speed to delivery	X	X
Tech transfer to established platforms	*X*	*X*
Faster to commercial image		X
Decrease bridging risks with early adoption	X	X
Minimal development for scale up	X	X
Synergy between disciplines	X	X
Process Safety		
Increased heat transfer	X	
Reduce inventory of energetic material	X	
Containment of hazardous materials	X	X
Minimize material in general	X	X

	DS	DP
Improved Control / Quality Assurance		
Real time analytical	X	X
Real time release	?	X
Reduction of unit operations	X	X
New tableting to minimize steps from API to finished product		X
Modeling of continuous processes	X	X
Modeling process disruptions	X	X
Minimize cross contamination	X	
Minimize unit operations	X	X
Manufacturing Efficiency		
Post approval change control strategies	X	X
Disposable equipment	X	
Reduce cycle time	X	X
Process intensification	X	X
Wider design space	X	X
mInimize material at risk	X	X
end to end CM	X	X
Supply Chain		
Minimize inventory	X	X
Speed	X	X

Barriers

	DS	DP
Lack of experience & uncertainty		
Resistance to doing something new	X	X
Competing cultures	X	X
Project attrition	X	X
Dose uncertaintly		X
Uncertainty in peak volume	X	
Containment level	X	X
Process validation strategies	X	X
Organizational acceptance of models	X	X
Control / Quality Assurance		
Complex control strategy	X	X
Cohesive materials. System accumulation	X	X
Start up waste	X	X
Diverting non conforming material	X	X
Analytical detection of low level impurities	X	
Traceability	X	X
Lifecycle of equipment	X	X

	DS	DP
Business Challenges		
What is the cost to go continuous?	X	X
Lack of infrastructure	X	X
Large existing batch infrastructure	X	X
Justifications for CM of low volume products	X	X
Underdeveloped external network	X	X
Lack of a powerful advocate	X	X
Multiple dev and manuf sites	X	X
Technical Challenges		
Feed rate limitations for low dose compunds		X
Lack of scale down models for development	X	X
Solids dosing for DS	X	
DS / DP interface for end to end CM	X	X
Molecular Complexity	X	
Underdeveloped external network	X	X

	DS	DP
Regulatory		
Regulatory buy in world wide	X	X
New technology in PAI	X	X
Fear of how to file what is a batch	X	X
What is sufficient verification / qualification for models	X	X

Major Discussion Themes

- Supply Chain Assurance
- CM implementation earlier in development
- Business case for CM
- Is end-to-end CM realistic?

Supply Chain Assurance

- Potential benefits include speed, flexibility, portability and inventory control.
- Barriers discussed included:
  - CMO capabilities for CM
  - Equipment lifecycle / Robustness
  - Internal capacity and equipment limitations

CM implementation earlier in development

- Benefits included reduced parallel efforts, avoid bridging concerns, consistent impurity control strategies and speed.
- Barriers
  - Lack of alignment at the interfaces. Requires education and improved interactions between disciplines.
  - Material availability. *Do more with less! Requires better scale down equipment and predictive models.*
  - Project timelines and uncertainty

Business case for CM

- Barriers:
  - Lack of detailed ROI analysis for demonstrated CM processes in pharma.
  - There are many publications on general economic advantages, but these lack the detail required to justify a real decision.
  - Sensitivity to publishing development costs.
  - Difficult to quantify many advertised CM advantages.
    - Value of improved quality assurance?

Opportunities

- Shared knowledge on defining the material properties that impact DP CM.
  - Many companies have individual efforts here.
  - Improved understanding could lead to earlier adoption, faster development and better process control.
- Push for publications on actual CM experiences.
  - True economic benefits, business cases, payback period, CapEx requirements, etc.
  - Examples of other CM advantages such as tech transfer, and improved QA.

Specific skills gaps

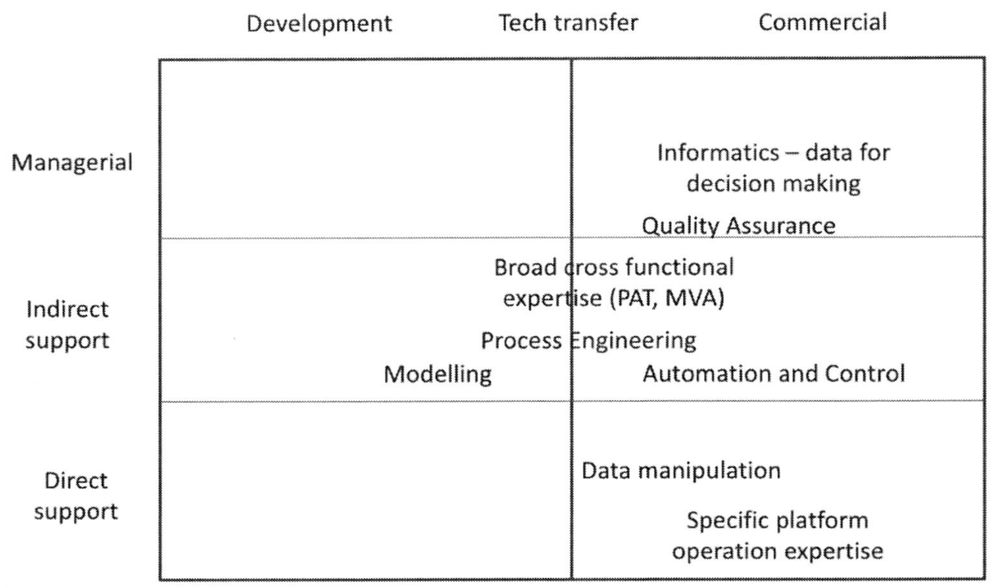

Mitigations

- Within company rotational assignments to develop broad multifunctional skills
- More use of academic consortium to develop and recruit skills
- Skills mapping across the development and manufacturing cycle
- FDA approach of web based modules built with academic and industry experts – expand access to industry
- Transferable skills from other sectors – do we value compliance experience over manufacturing paradigm experience; especially for automation and control
- Use of virtual training environments for operator level training

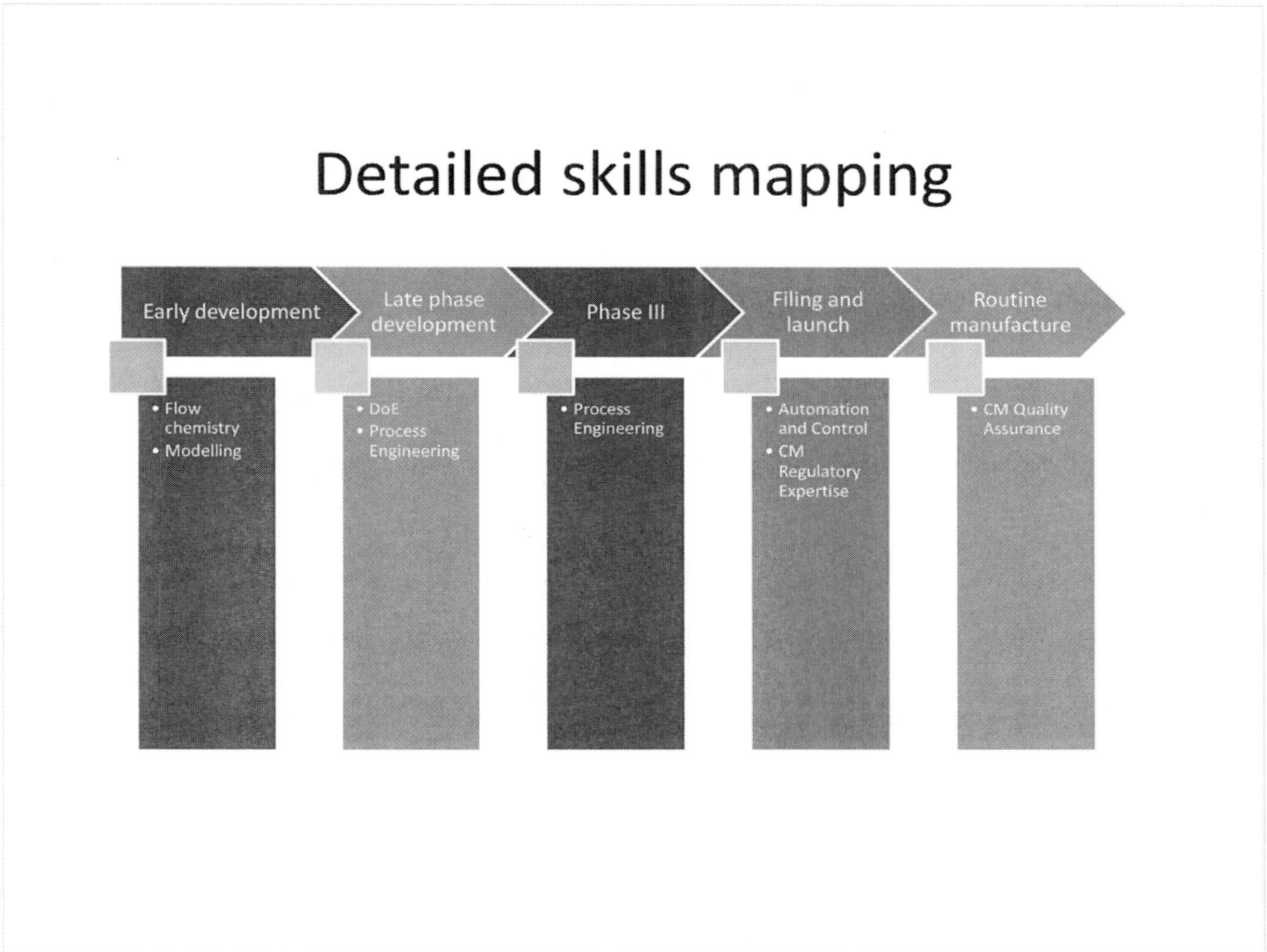
Detailed skills mapping
Early development
Late phase development
Phase III
Filing and launch
Routine manufacture
• Flow chemistry
• Modelling
• DoE
• Process Engineering
• Process Engineering
• Automation and Control
• CM Regulatory Expertise
• CM Quality Assurance

Topic 4: Needed Equipment and Associated Facility Cost Group Demographics (1)

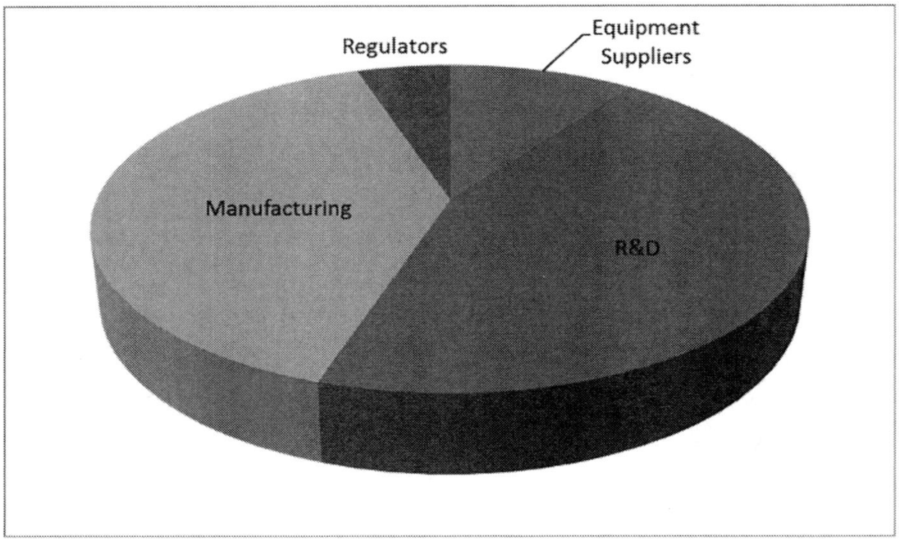

Topic 4: Needed Equipment and Associated Facility Cost Group Demographics (2)

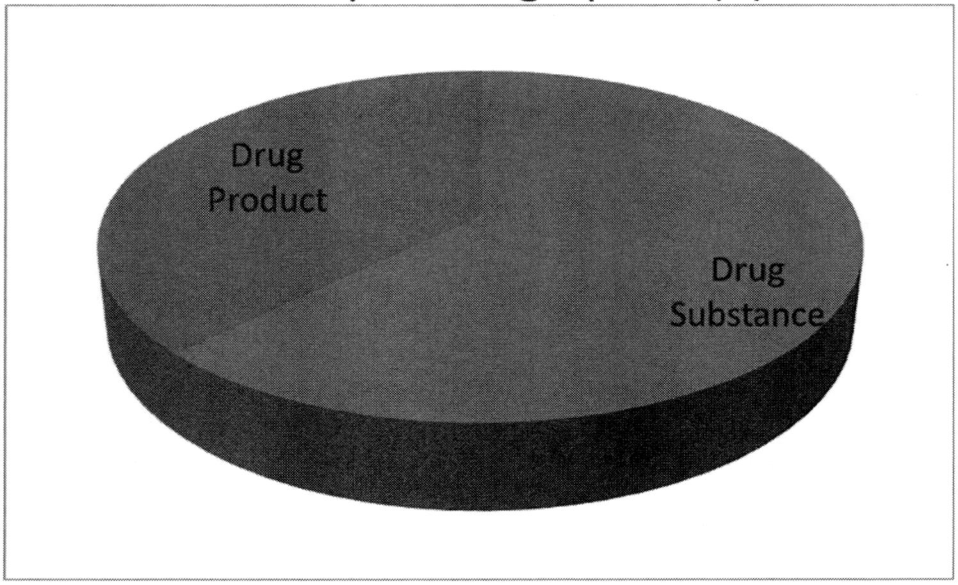

Statement 1

Recent investments in CP technology will spawn **iterative, innovation and prototyping of next generation CP equipment**, for use in the development, manufacture and supply of pharmaceuticals to the market place.

Recent investments in CP technology **will define how pharmaceuticals are developed, manufactured, and supplied** to the market for the decade(s) to come.

Statement 1

Recent investments in CP technology will spawn **iterative, innovation and prototyping of next generation CP equipment**, for use in the development, manufacture and supply of pharmaceuticals to the market place.

Recent investments in CP technology **will define how pharmaceuticals are developed, manufactured, and supplied** to the market for the decade(s) to come.

Statement 2

Emerging CP technologies **will be installed into or as part of existing bricks-and-mortar pharmaceutical manufacturing facilities**

Emerging CP technologies wil **radically and disruptively influence the design of pharmaceutical manufacturing facilities**

Statement 2

Emerging CP technologies **will be installed into or as part of existing bricks-and-mortar pharmaceutical manufacturing facilities**

Emerging CP technologies wil **radically and disruptively influence the design of pharmaceutical manufacturing facilities**

Topic 4: Needed Equipment and Associated Facility Cost

What are the perceived <u>technical</u> challenges / barriers for introducing new equipment?

- For new equipment, operational knowledge is often missing
  - Gaps will be filled over time with conferences and publications
  - Opportunity for collaboration between academia, industry and equipment provider
  - This relationship worked really well for PAT equipment
  - Appropriate scale down/up models of equipment

- Drug substance
  - Adapting PFRs to allow for heterogeneous reactions
  - Materials of construction
  - Continuous isolation and drying equipment
  - Quantitative charging of solids to equipment to remove need for solution make-ups

Topic 4: Needed Equipment and Associated Facility Cost

What are the perceived <u>technical</u> challenges / barriers for introducing new equipment?

- Drug Product
  - Design of excipients that aid flow (because APIs always flow well)
  - Mass feeders for the lower composition blend excipients
  - Containment, especially around feeding
  - Incorporation of ingredients for coating solutions
  - Relaxation period post tableting
- Analytical
  - Simple miniature sensors that behave and cost like thermocouples
  - Analysis for real-time content uniformity
  - Real-time chromatography
  - What do we do with all the data?
  - Be diligent with placement and selection of sensors – balance expensive sensors with other measurement methods

Topic 4: Needed Equipment and Associated Facility Cost

What are the perceived operational challenges / barriers for introducing new equipment?

- Cleanability
  - Opportunities to better engineer equipment with smaller number of parts
  - Difference in approach between DP and DS
- Ensuring containment, particularly around product transfers
- Training of support staff from operators to first-line technical support
  - Worked well for equipment suppliers to help with training even after the FATs
  - Perhaps even more exaggerated with CMOs
- Overcoming the activation barriers with all first-time introductions

Topic 4: Needed Equipment and Associated Facility Cost

What are the perceived facility challenges / barriers for introducing new equipment?

- Integration into the existing plant systems from Quality to Control
- Retrofit into existing brick and mortar structures
- Justifying the business case
- Unseating the batch king. Explaining that these are not like-for-like replacements, continuous equipment allows additional flexibility

Back up

Conference Program for the 2016 FDA-AIChE Workshop on Adopting Continuous Manufacturing

The Opportunity

Continuous processing has the potential to make the production of pharmaceuticals more responsive to patient needs. This workshop will address regulatory and other challenges associated with the implementation of this technology in pharmaceutical development and manufacturing. New technology conception to full incorporation has often required 20 + years. In some instances, uptake of technologies is never fully realized. Continuous manufacturing is the latest "initiative" that has met resistance to full-scale implementation despite the fact that many pharma companies used continuous manufacturing in the 1970's and 1980's for large-volume antibiotic and animal health products.

The Approach

This workshop addresses the issues surrounding full adoption of continuous manufacturing in pharmaceutical processing. Business cases for successful implementation will be shared, and focus groups will discuss these cases while addressing the top perceived issues. Technical sessions demonstrating the process and quality control benefits of implemented continuous manufacturing will be shared. The output of the workshop is targeted to inform ongoing regulatory guidance writing efforts.

Topic 4: Needed Equipment and Associated Facility Cost
Moderators: Dan Blackwood and Derek Berglund

- Folks will be able to post questions to our session ahead of time
- Dan and Derek to introduce ourselves
- Do we need a disclaimer? I assume to absolve everyone that they speak for themselves and not for their companies? AGREED
- Introduce goals of our session
  - be good stewards to encourage adoption of continuous processing technologies
  - Seek opportunities for pre-competitive co-design (Vendors, Pharma, Generics, Academics, etc) and commercialize CP equipment DS and 2nd generation DP
  - potential list of eqpt suppliers we should investigate NEEDs WORK

DO NOT PRESENT THIS SLIDE

Topic 4: Needed Equipment and Associated Facility Cost
Moderators: Dan Blackwood and Derek Berglund
Themes (4)

- What are the perceived benefits of equipment or facility standardization?

Topic 4: Needed Equipment and Associated Facility Cost
Moderators: Dan Blackwood and Derek Berglund
Themes (4)

- Are their benefits to standardizing equipment across the industry?
  - Would this reduce a PAI risk?
  - DP/DS?
  - Does this limit the eqpt to diversification in use?
  - What happens to vendor's IP position
  - How to encourage standardization? Push to ISPE/iQ/other symposia? Volunteers?
  - How can this be done without stifling innovation?
  - Opportunities for software standardization
- How to incentivize equipment manufacturers to develop prototype and then commercialize continuous equipment?
  - DP/DS/analytical
  - Can we garner any information from other industries on how to prototype and then pilot new equipment platforms?
  - What is the current state of equipment vendors? Do we have a comprehensive list?

Introduce 3 Provocative Statements

Recent investments in CP technology will define how pharmaceuticals are developed, manufactured, and supplied to the market for the decade(s) to come.	⟺	Recent investments in CP technology will spawn iterative, innovation and prototyping of next generation CP equipment, for use in the development, manufacture and supply of pharmaceuticals to the market place.
Emerging CP technologies will be installed into or as part of existing bricks-and-mortar pharmaceutical manufacturing facilities	⟺	Emerging CP technologies will radically and disruptively influence the design of pharmaceutical manufacturing facilities
The creation of CP platform technologies will drive the efficient adoption of CP paradigms into the pharmaceutical industry	⟺	The efficient adoption of CP paradigms into the pharmaceutical industry will constantly evolve through incremental and disruptive innovative of CP equipment design evolutions

Topic 4: Needed Equipment and Associated Facility Cost
Moderators: Dan Blackwood and Derek Berglund
Themes (1)

- What are the perceived technical challenges / barriers for introducing new equipment?
  - What is the largest need with respect to equipment?
    - Drug product / substance and analytical?
    - What is the current state of that equipment platform (prototype vs commercial ready)?
  - What unit operations are "shop-floor" ready for high performance operation
  - Equipment design for an uncertain future
  - What unit operations DP/API need additional design work
    - API: Reaction, Crystalization, Filtration, Drying, Milling
    - DP: Feeding, Powder Mixing, Granulation, Compression, Film Coating

Topic 4: Needed Equipment and Associated Facility Cost
Moderators: Dan Blackwood and Derek Berglund
Themes (2)

- What are the perceived operational challenges / barriers for introducing new equipment?
  - Resource and time requirements for completion of IQ/OQ/PQ activities
  - Training of workforce
  - Integrated CP equipment vs De-couple Batch processing equipment
    - Cleaning and Rapid change-over of a CP facility
    - Equipment Reliability to ensure high OEE
  - Implementation of Low Cost/Low Maintenance/High Impact advanced sensors
  - Complexity: doesn't fit with typical operations paradigm
  - How to integrate into existing systems? Cleaning? Space? Automation?

Topic 4: Needed Equipment and Associated Facility Cost
Moderators: Dan Blackwood and Derek Berglund
Themes (3)

- What are the perceived facility challenges / barriers for introducing new equipment?
  - Retrofit of existing facility to accommodate new integrated process equipment.
  - Coupling of material dispensing activities with material transformation/processing activities
  - Space requirements of CP equipment (e.g. headroom)
  - Facility design for an uncertain future?

Topic 4: Needed Equipment and Associated Facility Cost
Moderators: Dan Blackwood and Derek Berglund
Themes (4)

- Are their benefits to standardizing equipment across the industry?
  - Would this reduce a PAI risk?
  - DP/DS?
  - Does this limit the eqpt to diversification in use?
  - What happens to vendor's IP position
  - How to encourage standardization? Push to ISPE/iQ/other symposia? Volunteers?
  - How can this be done without stifling innovation?
  - Opportunities for software standardization
- How to incentivize equipment manufacturers to develop prototype and then commercialize continuous equipment?
  - DP/DS/analytical
  - Can we garner any information from other industries on how to prototype and then pilot new equipment platforms?
  - What is the current state of equipment vendors? Do we have a comprehensive list?

U.S. Food and Drug Administration
Protecting and Promoting Public Health
www.fda.gov

How Process Models can Facilitate Quality Risk Management for Emerging Technologies

Thomas O'Connor, Ph.D.

FDA-AIChE Workshop on Adopting Continuous Manufacturing
North Bethesda, MD
March 2th, 2016

U.S. Food and Drug Administration
Protecting and Promoting Public Health
www.fda.gov

Disclaimer

This presentation reflects the views of the authors and should not be construed to represent FDA's views or policies

2

U.S. Food and Drug Administration
Protecting and Promoting Public Health
www.fda.gov

Outline

- Initiatives to Support Adoption of Emerging Technologies
  - Emerging Technology Team
  - Regulatory Science Program
- Regulatory Science Initiatives Supporting CM
- Utilization of Models to Support Quality Risk Management
- Direct Compression Case Study
  - Risk Assessment
  - Risk Mitigation
  - Risk Communication
- Concluding Remarks

3

U.S. Food and Drug Administration
Protecting and Promoting Public Health
www.fda.gov

Drivers for Modernizing Pharmaceutical Manufacturing

- Advancing a 21st-Century Vision for Quality

  "A maximally efficient, agile, flexible pharmaceutical manufacturing sector that reliably produces high quality drugs without extensive regulatory oversight."

- Current Challenges Related to Product Quality

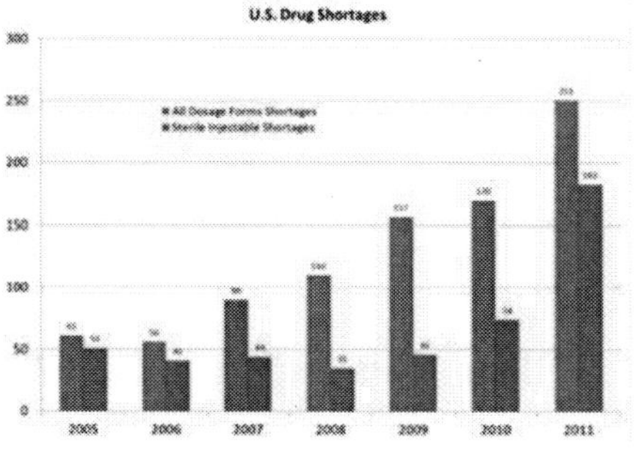

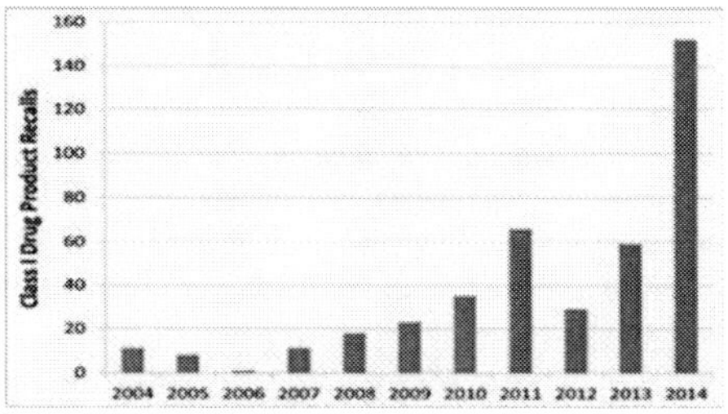

[1]U.S. Food and Drug Administration, "Drug Recalls," [Online]. Available: http://www.fda.gov/Drugs/DrugSafety/DrugRecalls/default.htm

4

Emerging Technology

- One of OPQ's stated goals is to *Encourage development and adoption of emerging pharmaceutical technology*
- What is an Emerging Technology?
  - Technology that has the potential to modernize the body of knowledge associated with pharmaceutical development to support more robust, predictable, and/or cost-effective processes or novel products and with which the FDA has limited review or inspection experiences, due to its relative novelty
- Examples of Emerging Technology include:
  - Novel methods of manufacturing
    - **Continuous manufacturing of drug substance and drug product**
    - "On-demand" manufacturing of drug products
    - Use of robots in pharmaceutical manufacturing
    - 3-D printed tablets
    - New container and closure system for injectable products

5

FDA Emerging Technology Team (ETT)

- Vision
  - Encourage and support the adoption of innovative technology to modernize pharmaceutical development and manufacturing where the Agency has limited review or inspection experience
- Objectives
  - To serve as a centralized location for external inquiries on novel technologies
  - To provide a forum for firms to engage in early dialog with FDA to support innovation
  - To ensure consistency, continuity and predictability in review and inspection by serving a lead role on the quality assessment team
  - To identify and evaluate roadblocks relating to existing guidance, policy, or practice
  - To help establish review and inspection standards and policy, as needed
  - Contact us: CDER-ETT@fda.hhs.gov
- ETT is a small cross functional team with representation from all relevant CDER and ORA review and inspection programs
- Draft Guidance for Industry: Advancement of Emerging Technology Application to Modernize the Pharmaceutical Base, December 2015

6

OPQ Science and Research Staff (SRS)

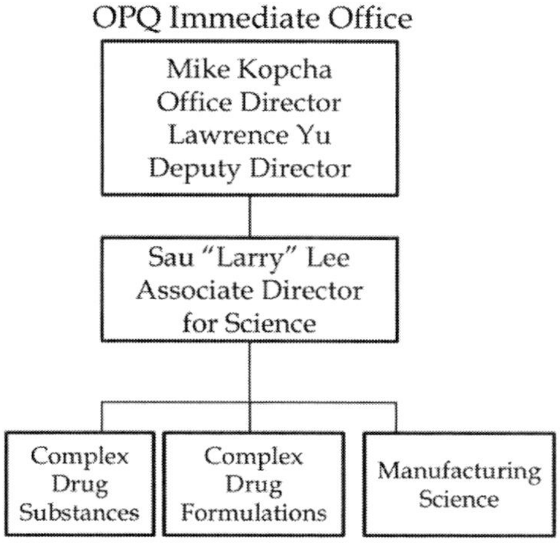

- Manufacturing Science Team
  - Manage external research projects (grants and contracts)
  - Develop and implement computational tools to support the risk-based assessment of drug manufacturing processes and quality
  - Interface with quality review teams through the ETT
  - Communication and workshop planning
  - Collaborates and coordinates with OTR & OBP in developing in-house advanced manufacturing lab programs
  - Consults on complex scientific issues with appropriate OPQ office(s)

7

U.S. Food and Drug Administration
Protecting and Promoting Public Health
www.fda.gov

Regulatory Science Initiatives Supporting CM

- Enhancing Regulatory Science for the Risk Based Assessment of Emerging Manufacturing Technologies Projects
  1. **Process simulation and modeling tools to facilitate the application of quality risk management**
  2. Investigations into material properties and equipment configurations; linking design space to final product quality
  3. Establishment of knowledge base to assess advanced process control & monitoring approaches (active controls, multivariate statistical process control, etc.)
  4. Concept development for utilizing process and materials data (raw and in-process) that support real-time release testing
- BARDA-FDA Continuous Manufacturing Innovations Initiative
  - Amendment to FDA's Advancing Regulatory Science BAA
- OTR and OBP developing in-house laboratory capability to enhance FDA's knowledge in manufacturing science to support CM implementation

8

U.S. Food and Drug Administration
Protecting and Promoting Public Health
www.fda.gov

Applications of Process Modeling

- Process models can provide a framework for process design, risk assessment, process control, and optimization
- Process models imbedded in Process Systems Engineering tools can provide benefits throughout the product lifecycle

Process Systems Engineering (PSE) Tools	Process Development Objectives
Predictive models	Process Understanding
Flowsheet modeling	Process Integration and Simulation
Feasibility analysis	Operating Space
Steady-state Optimization	Process Design
Dynamic Optimization	Process Efficiency
Sensitivity Analysis	Risk Assessment
Controller Design	Operations Excellence

9

U.S. Food and Drug Administration
Protecting and Promoting Public Health
www.fda.gov

Why Utilize PSE Tools

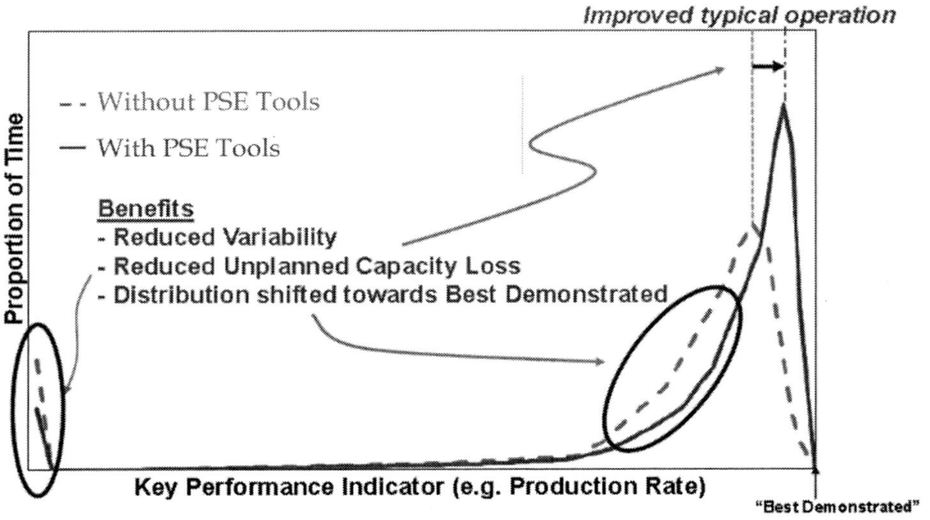

10

U.S. Food and Drug Administration
Protecting and Promoting Public Health
www.fda.gov

Continuous Manufacturing as a Potential Driving Force for PSE Tools

- Inherently data rich processes
- Availability of plant wide information systems
- Advancements in process modeling and simulation

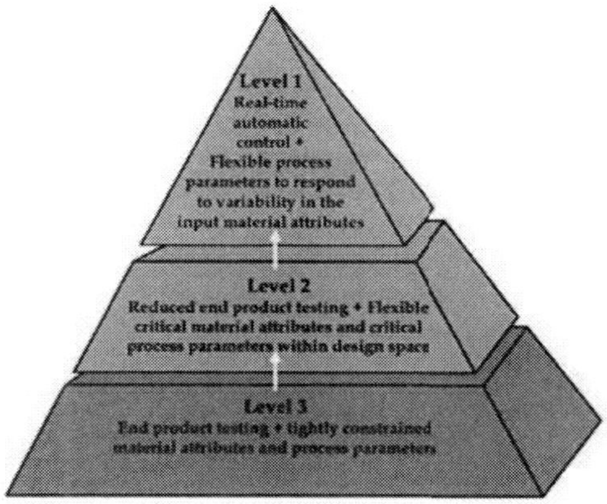

Lee S. *et. al.* J Pharm Innov. **2015** DOI 10.1007/s

Many continuous manufacturing systems promote the adoption of higher level controls, although a hybrid approach combing the different levels of control is viable for some product and process designs

11

U.S. Food and Drug Administration
Protecting and Promoting Public Health
www.fda.gov

Development of Process Simulation and Modeling Platform

- Utilization of models in a regulatory setting is not new
  - Pharmacokinetics and drug absorption modeling (e.g., GastroPlus simulations)
- However, process modeling of pharmaceutical powder processes is not yet mature and requires further development
  - Advancing the development of unit operation model libraries
  - Integrated process models (i.e. flowsheet models) including process controls
  - Post-processing tools to a facilitate risk assessments
- FDA issued two grants for the development of process simulation and modeling tools for integrated pharmaceutical manufacturing processes
  - Rutgers University
  - University of Massachusetts at Lowell
- Aim of the program is to create a collaborative platform for process simulation
  - Build on process modeling knowledge developed in academia, industry, equipment vendors and regulatory bodies

12

Process Models as an Emerging Tool to Support Quality Risk Management

- Risk assessment: Models can be utilized to identifying hazards and failure modes
  - Process understanding forms the foundation for effective risk management
- Risk mitigation: Models can be utilized as part of the control and/or used to evaluate the effectiveness of the control strategy
  - Model based control, multivariate monitoring, analysis of large of data sets, and/or Real Time Release Testing (RTRT)
- Risk communication: Process models can be good framework for capturing the related scientific knowledge and communicate residual levels of risk
  - Additional process understanding gained and communicated throughout the lifecycle of the drug product could be utilized as part of continual improvement of the process
- Process models can support the effective implementation of emerging technologies
- Process modeling is one of the approaches to the meet the general expectation of a science and risk-based understanding and control of processes and product quality

13

Direct Compression Case Study to Illustrate Research Initiatives

- Many sources of variability in continuous manufacturing process
  - Raw material properties
  - Design parameters
  - Operating conditions
- Applications of sensitivity analysis:
  - Identification of critical process parameters
  - Identification of potential control variables and development of control strategies

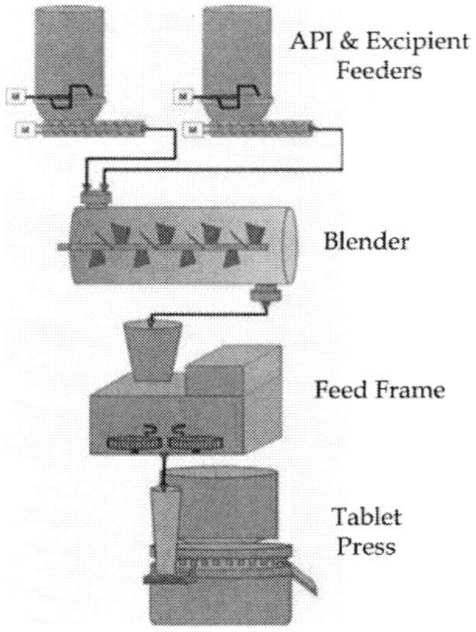

14

Sensitivity Analysis and Risk Assessment

Model Based Risk Assessment for Continuous Process

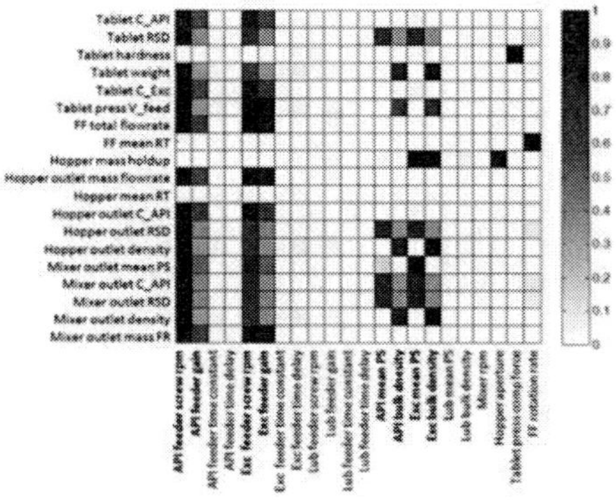

Current Example of an Initial Risk Assessment Batch Process for IR Solid Oral Drug Product

PRODUCT PROPERTY/IMPACT OF CHANGE/CQAS	PROBABILITY OF OCCURRENCE (O)	SEVERITY OF EFFECT (S)	DETECTABILITY (D)	FMECA RPN	Comment
Physical stability (solid state)	Crystalline (3)	1	4	12	BCS class I
Chemical stability	Increasing trend (4)	3	4	48	Increasing trend of Related Compound A
Assay	2	3	3	18	
Content uniformity	>10% (2)	2	4	16	77.8% drug load, wet granulation
Microbial limits	1	3	3	9	
Dissolution IR	3	2	2	12	

Preliminary screening

- Focus subsequent assessment on most significant parameters
- Feeder parameters - Screw speed; Feed factor

Boukouvala *et. al.* Comput. Chem. Eng. 2012;42;30-47
Lakerveld R *et. al.* AIChE J. 2013;59;3671-3685

Process Understanding: Dynamics

- Understanding the interactions of unit operations in the process train help understand the behavior of the entire system (over time)
- Residence Time Distribution (RTD) depends material properties, process parameters, & equipment configuration

Line rate is a important variable to be considered

RTD as a function of back-mixing

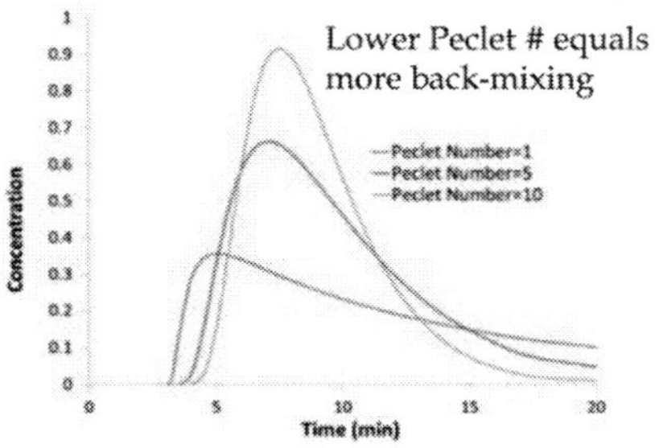

RTD as a function of line rate

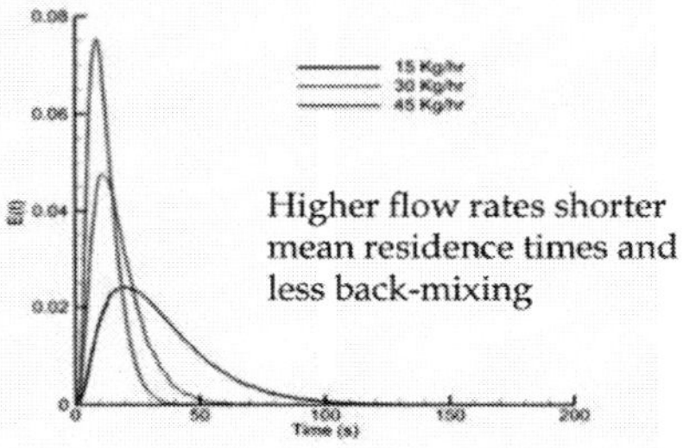

U.S. Food and Drug Administration
Protecting and Promoting Public Health
www.fda.gov

Process Dynamics – Detecting Disturbances

- Detectability of transient disturbances impacted by relationship between process dynamics and sampling frequency

API concentration pulses resulting in OOS material

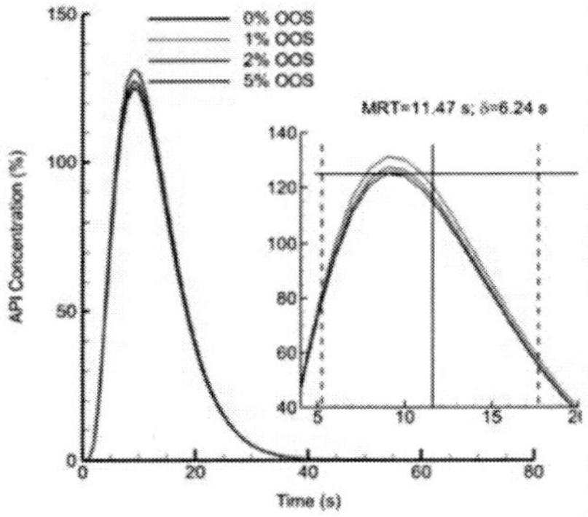

Impact of RTD on Detectability

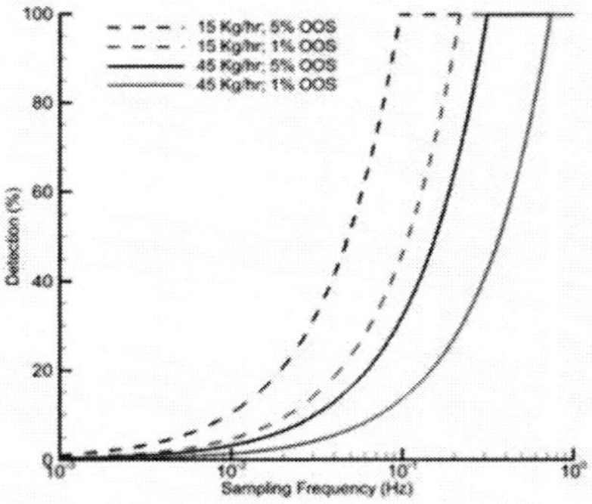

Engisch W. and Muzzio F.. J Pharm Innov. 2015; DOI 10.1007/s12247-015-9238-1

17

U.S. Food and Drug Administration
Protecting and Promoting Public Health
www.fda.gov

Process Dynamics – Formulation Variability

- Feeding variations of excipients may impact product performance (e.g. dissolution)

Feeding Variability

Tablet Variability

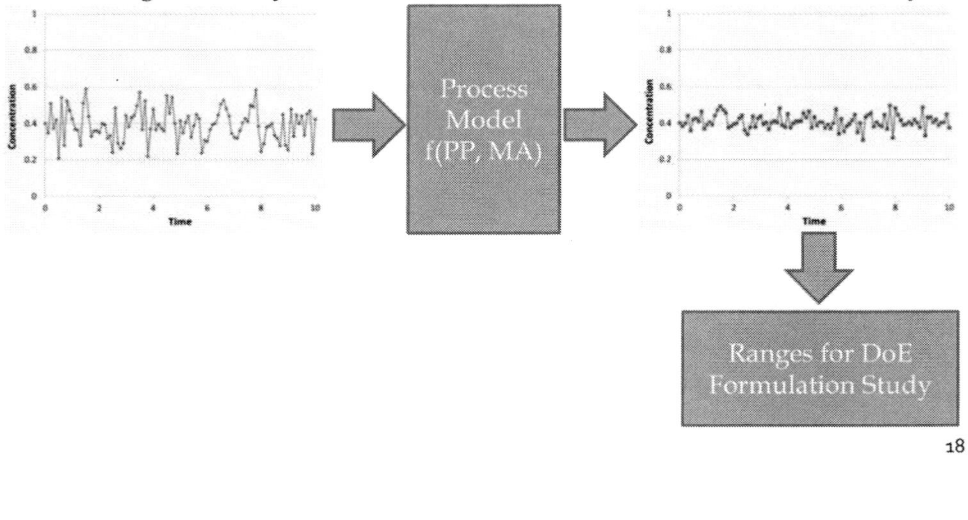

18

Risk Mitigation - Advanced Control Approaches

- Control Strategy approaches can be incorporated into the flowsheet model
- Ratio Control: used to maintain the flow rate of one stream at a specified proportion relative to another flow rate
- Feedback: output information (controlled variable) is used to automatically trigger upstream action (manipulated variable)
  - Typically incorporated into the control of individual unit operation (e.g. loss-in-weight feed rate control, compression force control)

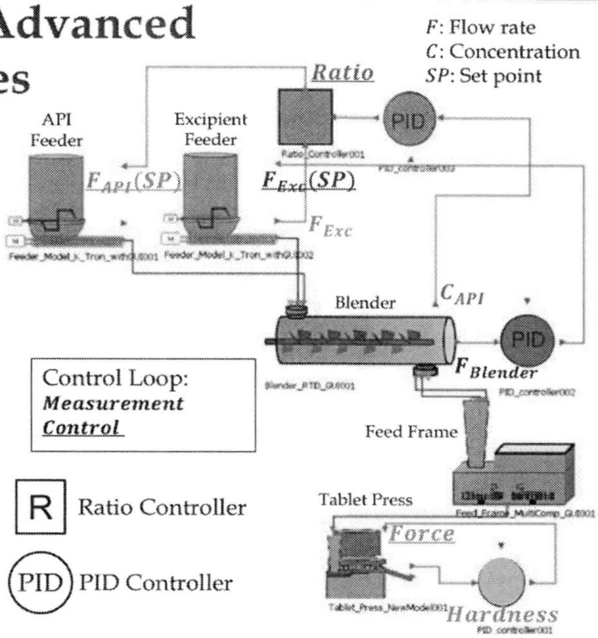

Case Study Evaluating Control Strategy

- Time with out-of-spec API concentration: API feeder disturbance

Without Controller

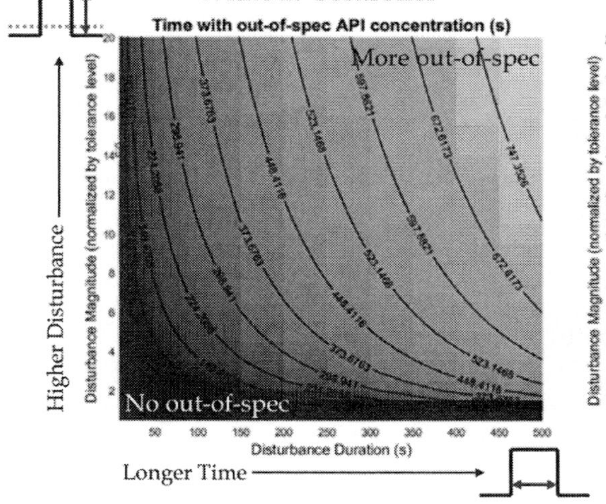

With Controller

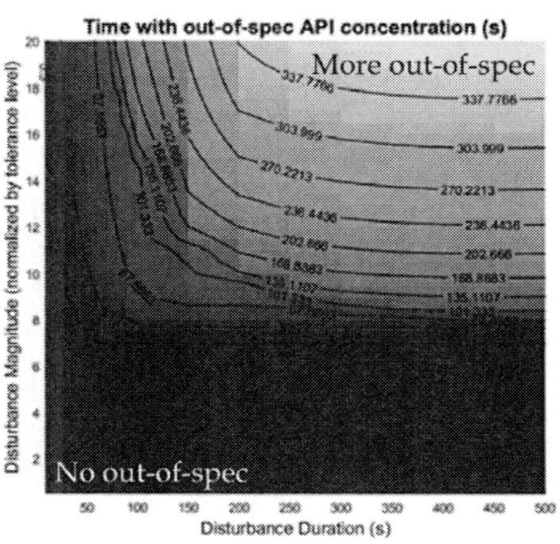

Risk Mitigation – Utilizing Models to Divert of Nonconforming Material

- The evaluation of overall residence time distribution and the understanding of propagation of a disturbance between extraction points in the system are important to justify the amount of material at risk due to an unexpected even or disturbance
  - Analysis should account for uncertainty in the measured/predicted RTD

Tank in Series Model for RTD

$$E(t) = \frac{t^{n-1}}{(n-1)!\left(\frac{\tau}{n}\right)^{n}} e^{\left(\frac{-nt}{\tau}\right)}$$

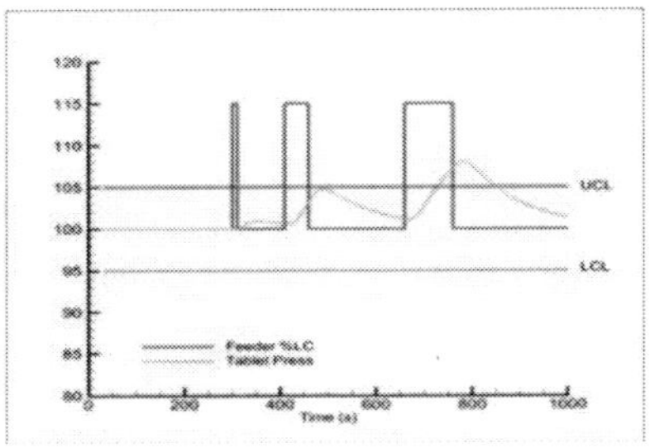

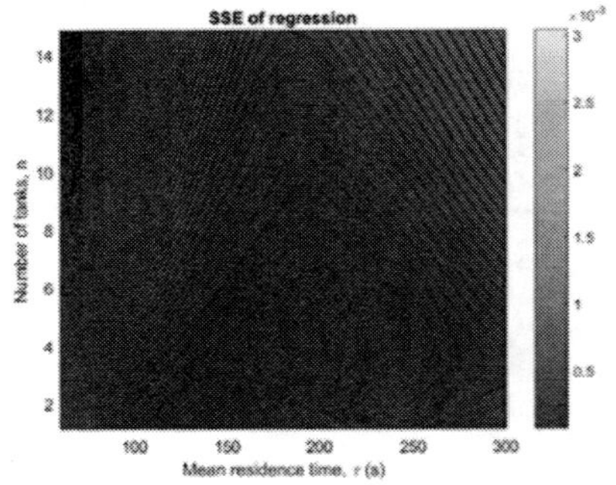

2-D contour of minimized SSE for RTD model

Scientific and Risk-Based Considerations for Model Based Material Diversion

- Develop the model utilizing scientifically sound principles and conditions that directly reflect routine commercial production
- Verify performance for high-impact models
  - E.g. Capability of the model to trace the identified OOS product segment through the system to the rejection point
- Understand risks to validity of model predictions
  - Model parameter uncertainty
  - Expected variation in process parameters and material attributes *including line rate*
  - Assessment of product quality risks resulting from potential transient disturbances and/or process failure modes that may not be identified by or included in the model
- Include model maintenance approaches within the quality system as part of a lifecycle approach
  - Routine monitoring to verify performance

U.S. Food and Drug Administration
Protecting and Promoting Public Health
www.fda.gov

Risk Communication: Process Models

- Quality risk management requires product and process knowledge to identify, analyze, evaluate, control, and communicate the risks
- Process models can be good framework for capturing the related scientific knowledge
  - Formally stating assumptions and reasonable sources of uncertainty enhances confidence in this output and/or helps identify its limitations
  - Utilization of process models in risk-based quality management can become one source of common risk assessment approaches, and can be used to facilitate the communication of risk and risk mitigation approaches between industry and regulatory bodies

23

U.S. Food and Drug Administration
Protecting and Promoting Public Health
www.fda.gov

Concluding Thoughts

- OPQ supports the modernization of manufacturing in part through the ETT and the advancement of regulatory science
- Continuous manufacturing is a potential driving force for utilization of process model throughout a product lifecycle
- FDA has several ongoing research initiatives on advancing process modeling related to CM
- Process models are an emerging tool to support systematic risk management for pharmaceutical processes
- Interested in developing collaborations to address barriers

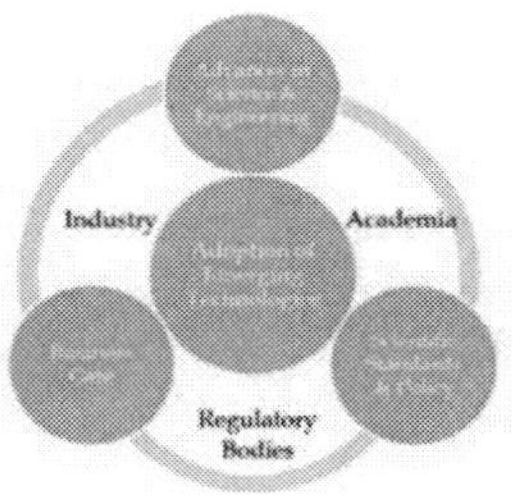

24

U.S. Food and Drug Administration
Protecting and Promoting Public Health
www.fda.gov

Acknowledgements

OPQ Science and Research Staff

- Geng (Michael) Tian
- Xiaochuan (Ben) Yang

Emerging Technology Team

- Chair: Sau (Larry) Lee, Associate Director of Science, OPQ
- PM: Cheryl Kaiser (OPQ/OPRO)
- Members: Thomas O'Connor(OPQ/IO-SRS), Celia Cruz (OPQ/OTR), Mohan Sapru & Ray Frankewich (OPQ/ONDP), Geoffrey Wu (OPQ/OLDP), Kurt Brorson (OPQ/OBP), Grace McNally, Sharmista Chatterjee & Bryan Riley (OPQ/OPF), Rebeca Rodriguez & Susanne Richardson (ORA)

External Collaborators

- Rutgers University
- UMass Lowell
- PSE
- MIT

25

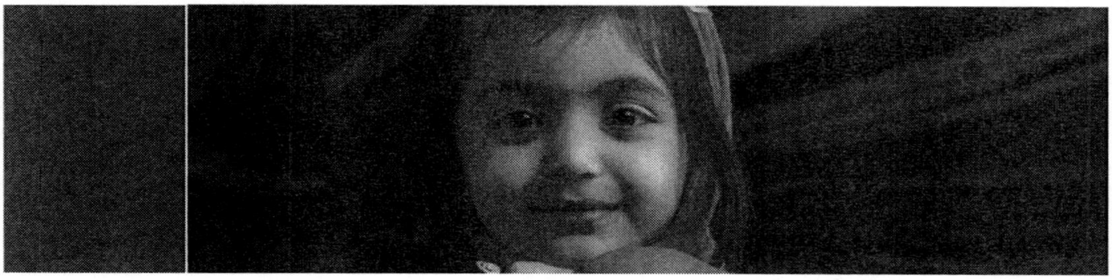

Strategies to Accelerate Process Control Adoption in the Pharmaceutical Industry

Richard D. Braatz

Outline

1. Model-Based Plantwide Control
2. Strategies from the May 20–21, 2014 Continuous Manufacturing Symposium
3. Strategies partly from the October 19, 2015 Promoting Continuous Manufacturing Workshop

Integrated Control Strategy for Continuous Manufacturing

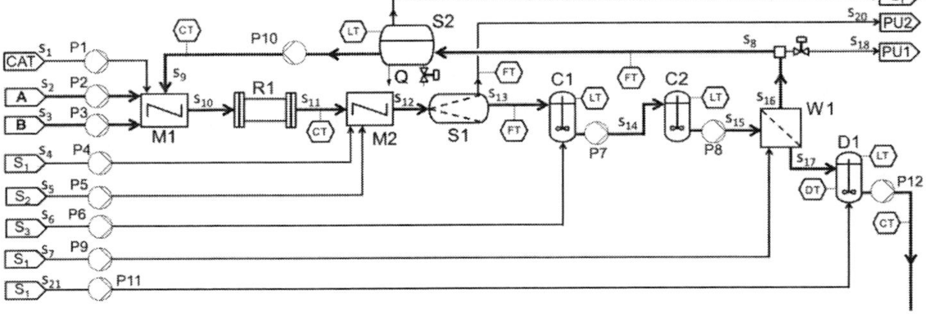

- Tight integration of operations can result in disturbances propagating downstream, unless their effects are suppressed by an integrated control strategy
- The control strategy must optimize the overall plant operation instead of only isolated units (i.e., need *plantwide control*)

Plantwide Control of Continuous Manufacturing

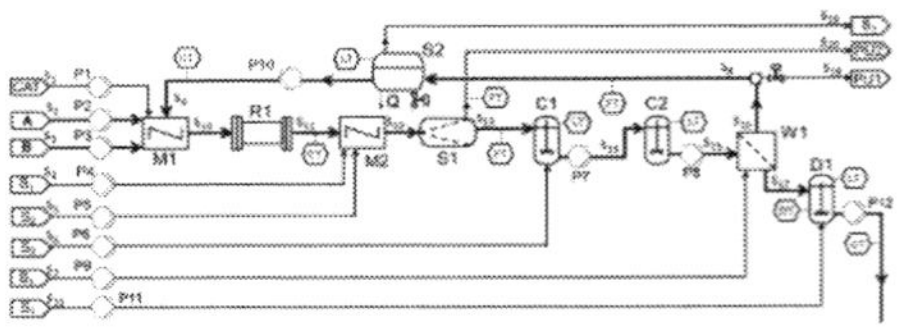

- System characteristics
  - Many connected unit operations
  - Very fast to slow processes
  - Multi-purpose plants with short development time
  - Alignment with regulatory requirements
- Approach adapted from the chemical industry
  - Employ systematic and modular design of plantwide control strategies for continuous manufacturing facilities
  - Experimentally demonstrated on continuous pilot plant

R. Lakerveld, B. Benyahia, P.L. Heider, H. Zhang, A. Wolfe, C. Testa, S. Ogden, D.R. Hersey, S. Mascia, J.M.B. Evans, R.D. Braatz, and P.I. Barton. The application of an automated control strategy for an integrated continuous pharmaceutical pilot plant. *Organic Process Research & Development*, 19, 1088-1100, 2015

A continuous pilot plant

S. Mascia, P.L. Heider, H. Zhang, R. Lakerveld, B. Benyahia, P.I. Barton, R.D. Braatz, C.L. Cooney, J.M.B. Evans, T.F. Jamison, K.F. Jensen, A.S. Myerson, and B.L. Trout. End-to-end continuous manufacturing of pharmaceuticals: Integrated synthesis, purification, and final dosage formation. *Angewandte Chemie*, 52(47):12359-12363, 2013

Model-based Plantwide Control

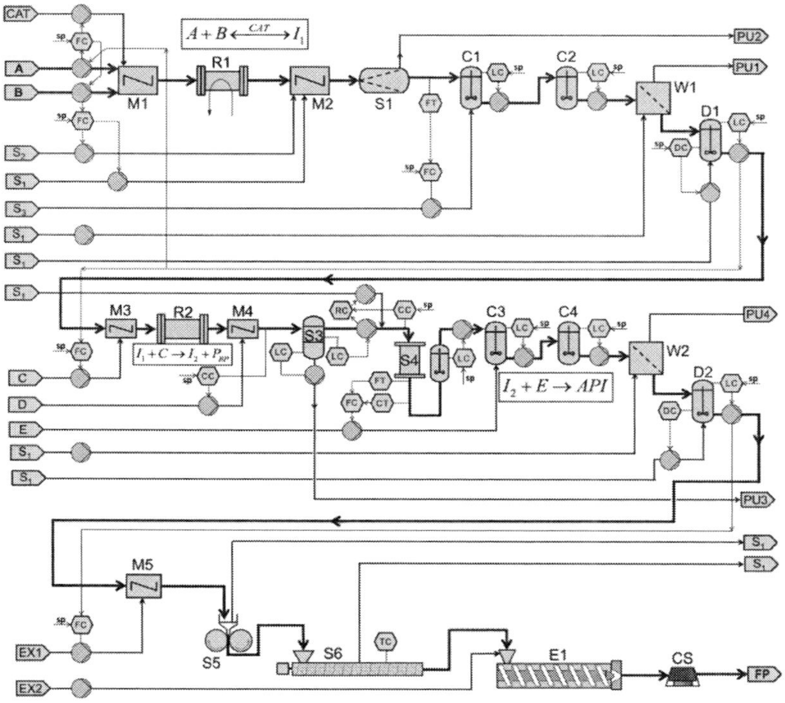

- First-principles dynamic models were built for each unit operation (UO) as they were developed
- Models were validated and then placed into a plant-wide simulation (with uncertainties)
- Plant simulation used to design plantwide control strategy

R. Lakerveld, B. Benyahia, ..., R.D. Braatz, and P.I. Barton. *Organic Process Research & Development*, 19, 1088-1100, 2015

Model-based Plantwide Control

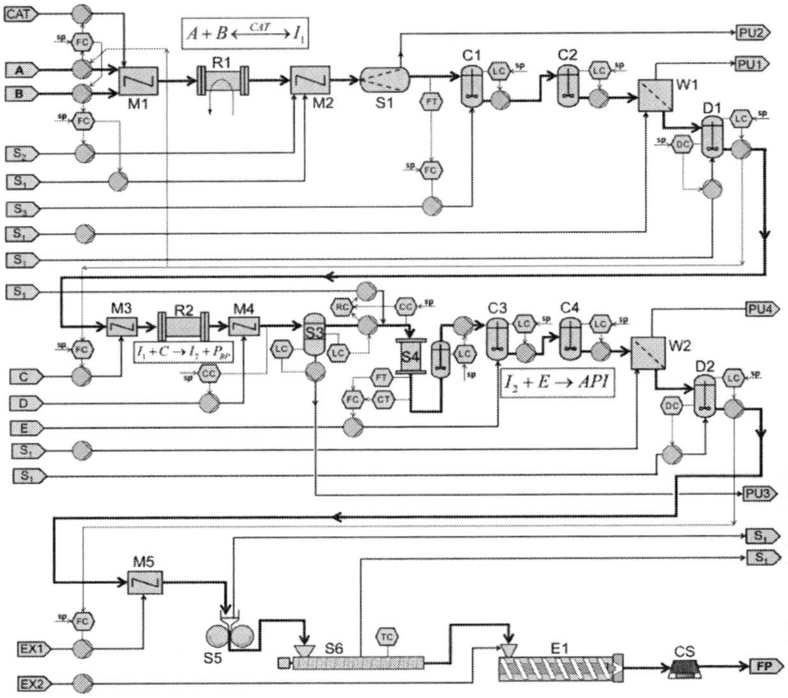

Met all purity specs in Summer 2012

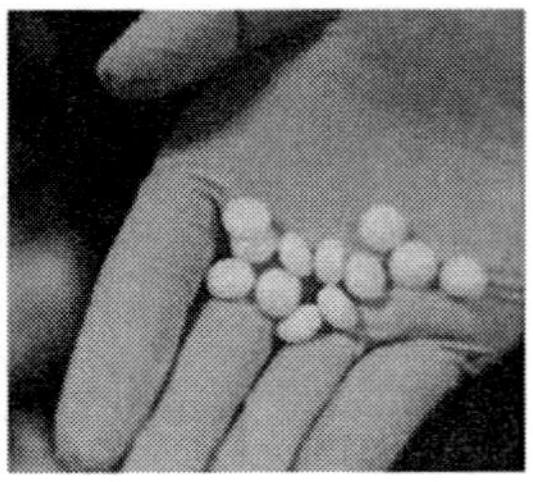

S. Mascia, P.L. Heider, H. Zhang, R. Lakerveld, ..., A.S. Myerson, and B.L. Trout. *Angewandte Chemie*, 52, 12359-12363, 2013

Outline

1. Model-Based Plantwide Control
2. <u>Strategies from the May 20–21, 2014 Continuous Manufacturing Symposium</u>
3. Strategies partly from the October 19, 2015 Promoting Continuous Manufacturing Workshop

Key Control-related Messages (from the May 20–21, 2014 Continuous Manufacturing Symposium)

- The main objective of continuous operations should be on being "in control" rather than being at steady state
- Continuous operations require a plant-wide control strategy that ensures that all CQAs are satisfied
- A monitoring system needs to track material as it moves through the manufacturing facility, using PAT and RTDs
- Systematic approaches are needed to manage constraints, disturbances, nonlinearities, uncertainties, and risk
- Regulatory bodies need to ensure that regulations and regulatory practices promote and do not derail continuous

Allan S. Myerson, Markus Krumme, Moheb Nasr, Hayden Thomas, and Richard D. Braatz. Control Systems Engineering in Continuous Pharmaceutical Manufacturing. *J. Pharm. Sci.*, 104, 832–839, 2015

MIT NOVARTIS

Special Control-related Technical Needs (from the 2014 CM Symposium)

1) Design of optimal startup and shutdown procedures
2) Design of process monitoring and control systems that collectively provide high quality assurance
3) Control strategies for specific new unit operations
4) Development of systems integration methods that respect the higher quality assurance needed in pharmaceuticals
5) Understanding the integration of design spaces and quality assurance with design of overall plant-wide control strategy
6) Design/tuning of control systems for each unit operation to take into account disturbances, nonlinearities, dynamics, constraints, and uncertainties
7) The quantification of the technical risks of failures or delays that occur anywhere in process development

MIT NOVARTIS

Suggested strategies for accelerating widespread adoption of process control (from the CM Symposium)

- Generate more graduates with expertise in pharmaceutical process control
  - Federal support for pharmaceutical manufacturing is very low in most countries, and university pharma centers rarely hire control engineers
  - Can industry and regulatory bodies work together with other federal agencies to create funding mechanisms that are competitive with high-money areas like biomedical engineering?
- Speed the correction of misconceptions about continuous pharmaceutical process control/operations, e.g.,
  - Steady state vs. "in control"
  - Artificially defined batches vs. residence time distribution functions
  - Understanding the value of feedback vs. feedforward control

Suggested strategies for accelerating widespread adoption of process control (from the CM Symposium)

- Regulatory bodies have a challenge in training the individuals who directly interact with companies on filings
  - Could financially support the development of high quality training materials in pharmaceutical process control for joint use by students, company employees, and regulatory staff
  - Could enhance training of their technical staff by supporting joint research projects with universities in continuous manufacturing and the associated process control technology
  - Could encourage other federal agencies to support research in pharmaceutical process control

Allan S. Myerson, Markus Krumme, Moheb Nasr, Hayden Thomas, and Richard D. Braatz. Control Systems Engineering in Continuous Pharmaceutical Manufacturing. *J. Pharm Sci.*, 104, 832–839, 2015

Outline

1. Model-Based Plantwide Control
2. Strategies from the May 20–21, 2014 Continuous Manufacturing Symposium
3. Strategies partly from the October 19, 2015 Promoting Continuous Manufacturing Workshop

MIT NOVARTIS

Suggested strategies for accelerating widespread adoption of process control

- Develop first-principles models and simulation techniques to support robust process design and control*
- Consider developing a modeling toolbox in a collaboration among government agencies, academia, and industry; then make available to stakeholders*
- Develop integrated automated control systems to facilitate straightforward deployment**

* *Promoting Continuous Manufacturing in the Pharmaceutical Sector*, The Brookings Institution, Washington, DC, October 19, 2015, http://www.brookings.edu/~/media/events/2015/10/19-continuous-manufacturing-pharmaceutical-sector/meetingsummary_101915_continuousmanufacturing.pdf

** Z.K. Nagy and R.D. Braatz. Advances and new directions in crystallization control. *Annu. Rev. Chem. Biomol. Eng.*, 3, 55–75, 2012

MIT NOVARTIS

Extra slide

Approach takes into account the four strategies for ensuring that a particular CQA spec is satisfied

1. Direct measurement of the CQA
2. Prediction of the CQA based on a first-principles model that is fed measurements of related variables and is running in parallel with operations
3. Prediction of the CQA based on an empirical or semi-empirical model (e.g., response surface map, PLS model) that is fed measurements of other variables

(1–3:) Can use in closed-loop feedback control strategy

4. Operation of the CPPs to lie within a design space, that is, some specified set shown in offline studies to provide assurance

(4:) Open-loop strategy

The smaller the number, the higher the quality assurance

An integrated multi-layer approach to control strategy for continuous drug product manufacture

Ahmad Almaya, Ph.D.
Eli Lilly and Company
February 2016

Lilly's Journey with DP CM

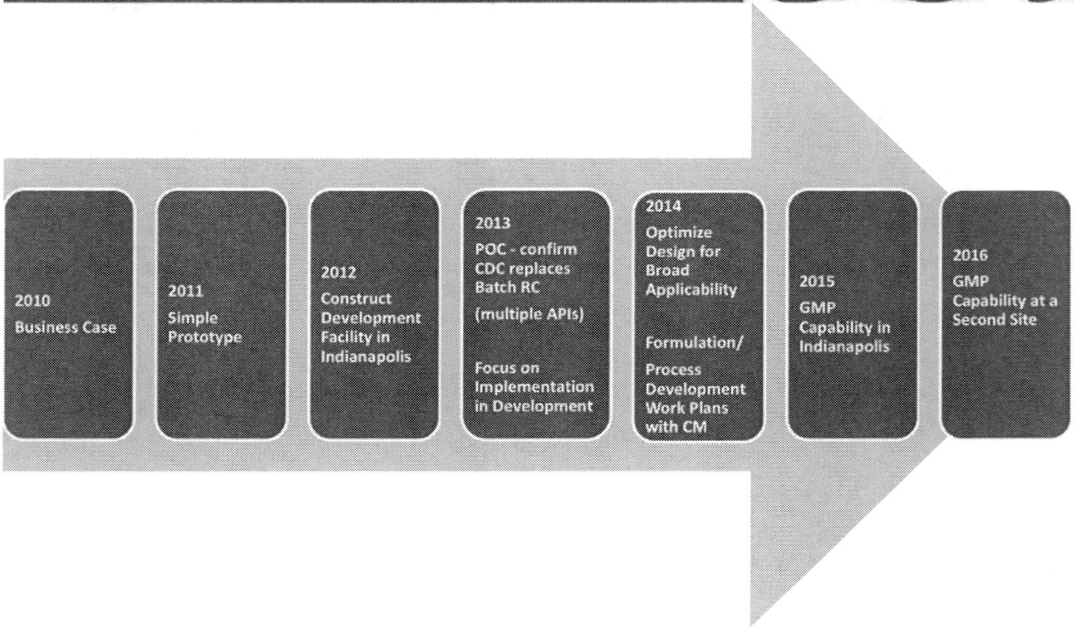

Lilly's Journey with DP CM

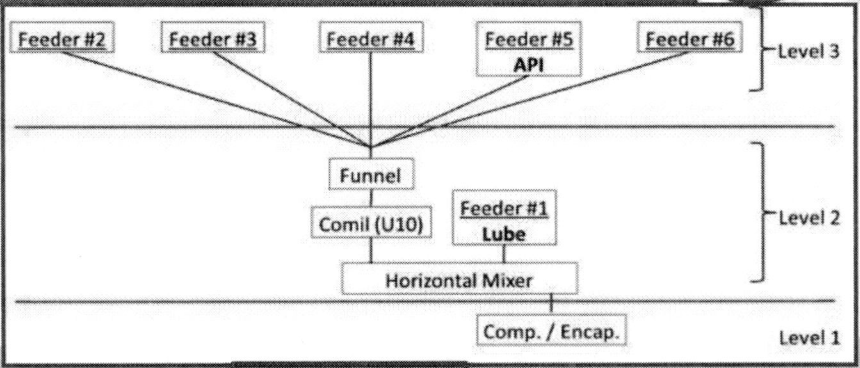

2010 Business Case

2011 Simple Prototype

2012 Construct Development Facility in Indianapolis

2013 POC - confirm CDC replaces Batch RC (multiple APIs) Focus on Implementation in Development

2014 Optimize Design for Broad Applicability Formulation/ Process Development Work Plans with CM

2015 GMP Capability in Indianapolis

2016 GMP Capability at a Second Site

Lilly's Journey with DP CM

Compound	Compound 1	Compound 2	Compound 3	Compound 4	Compound 5	Compound 6
Risks/ properties	Flowable API	Segregation Risk	Poor Flow (tablet weight control)	Capsules	Segregation Risk	Micronized DS, Flow/Uniformity Risks
Batch process	DC	RC	RC	Blend/Fill	RC	DC
CM process	DC	DC	DC	Blend/Fill	DC	DC
CU Pass/Fail	P	P	P	P	P	P

2010 Business Case

2011 Simple Prototype

2012 Construct Development Facility in Indianapolis

2013 POC - confirm CDC replaces Batch RC (multiple APIs) Focus on Implementation in Development

2014 Optimize Design for Broad Applicability Formulation /Process Development Work Plans with CM

2015 GMP Capability in Indianapolis

2016 GMP Capability at a Second Site

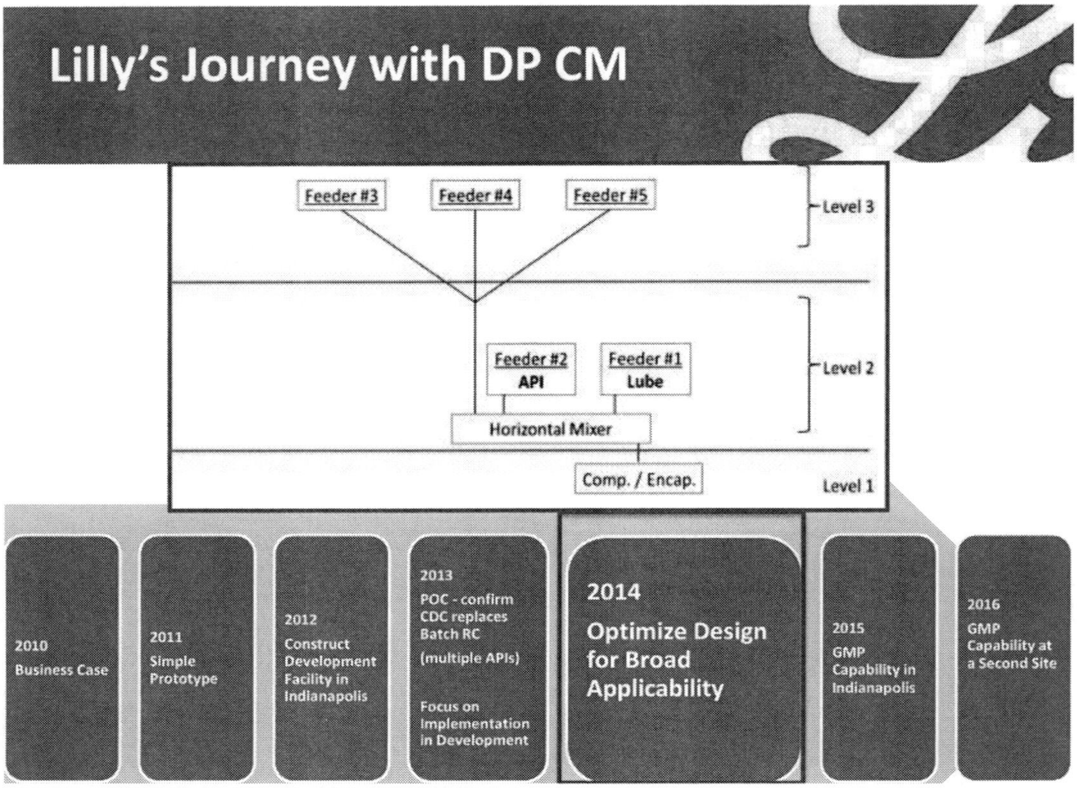
Lilly's Journey with DP CM
Feeder #3
Feeder #4
Feeder #5
Level 3
Feeder #2
API
Feeder #1
Lube
Level 2
Horizontal Mixer
Comp. / Encap.
Level 1
2010
Business Case
2011
Simple Prototype
2012
Construct Development Facility in Indianapolis
2013
POC - confirm CDC replaces Batch RC
(multiple APIs)
Focus on Implementation in Development
2014
Optimize Design for Broad Applicability
2015
GMP Capability in Indianapolis
2016
GMP Capability at a Second Site

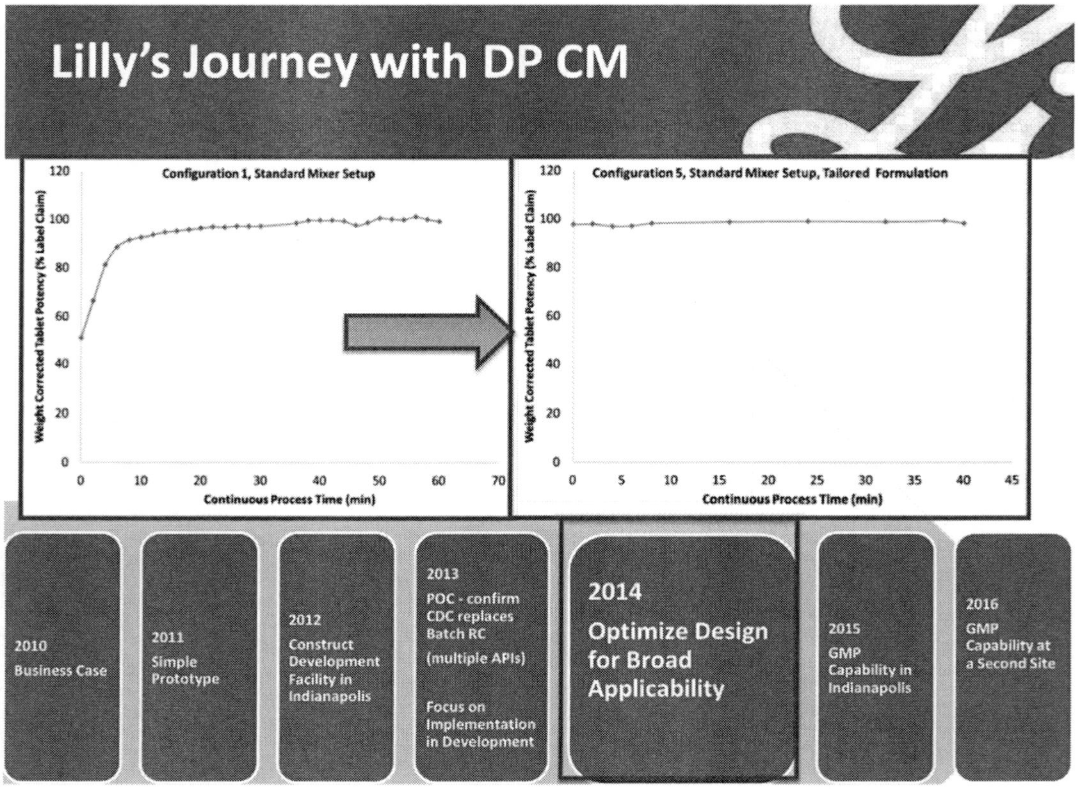
Lilly's Journey with DP CM
Configuration 1, Standard Mixer Setup
Weight Corrected Tablet Potency (% Label Claim)
Continuous Process Time (min)
Configuration 5, Standard Mixer Setup, Tailored Formulation
Weight Corrected Tablet Potency (% Label Claim)
Continuous Process Time (min)
2010
Business Case
2011
Simple Prototype
2012
Construct Development Facility in Indianapolis
2013
POC - confirm CDC replaces Batch RC
(multiple APIs)
Focus on Implementation in Development
2014
Optimize Design for Broad Applicability
2015
GMP Capability in Indianapolis
2016
GMP Capability at a Second Site

Lilly's Journey with DP CM

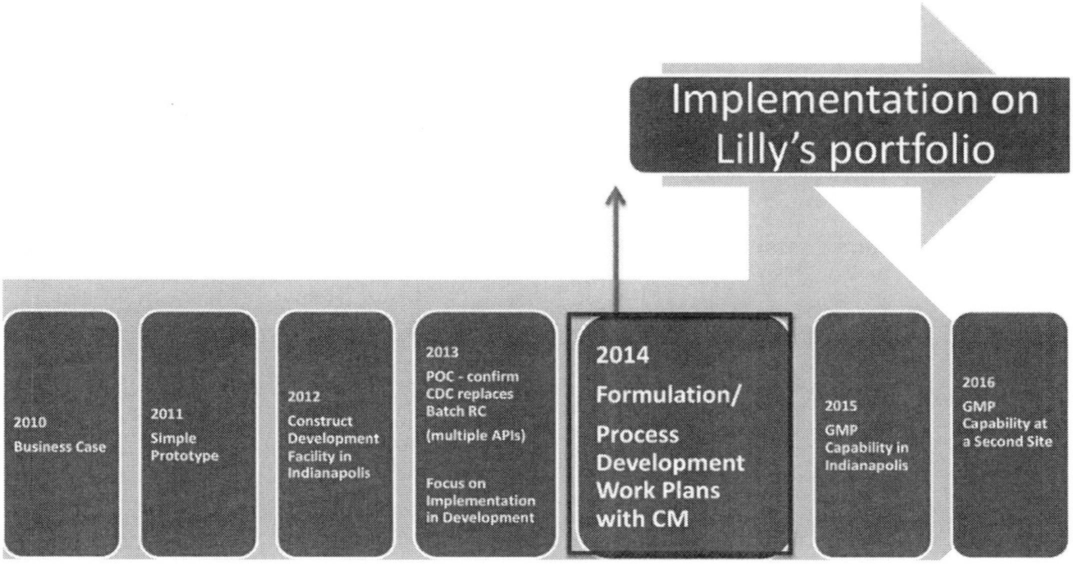

Optimized Design

- Unit Utilizes a DeltaV control system, SynTQ PAT controller and data synchronization, and CAMO Unscrambler multivariate analysis (MVA).
- GMP units at Lilly's manufacturing sites have identical equipment, PAT instrumentation and automation to the development unit.

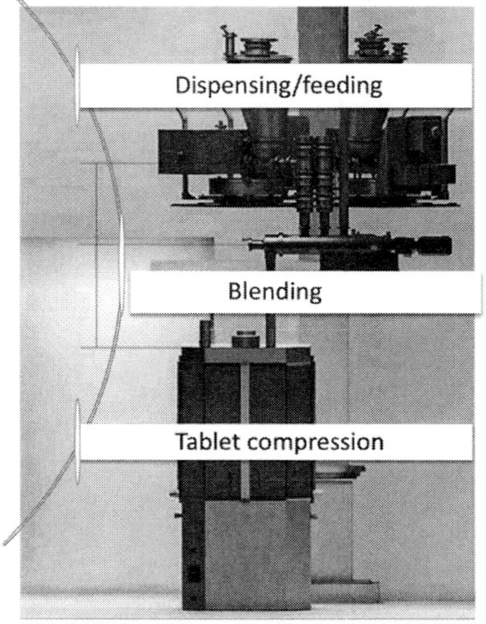

Advantages of Equipment Design Simplification

- Optimized system performance
  - Minimize special cause disturbances related to material flow (i.e., remove potential material accumulation areas and elements that negatively impact consistent material flow).
  - Improve startup phase (minimize startup waste).
- Enabling a robust control strategy
  - Minimize risk of potency and uniformity disturbances by real-time control of feeders, and robust mixer design and operation.
- Allow implementation of automation as part of the control strategy
  - Real time process monitoring and automated product collection/rejection

9

Operating Control Strategy - Feeders

LIW feeder monitoring and control are the primary elements to ensure drug product CQAs

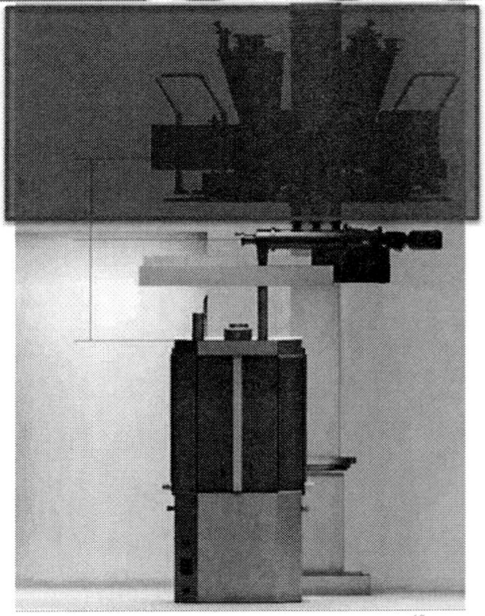

10

LIW Feeder – Ratio Mode

- Cascade control loop where the actual output from one feeder (e.g. the worst flowing material which is typically the drug substance) is used to calculate the flow rate set points for the other feeders

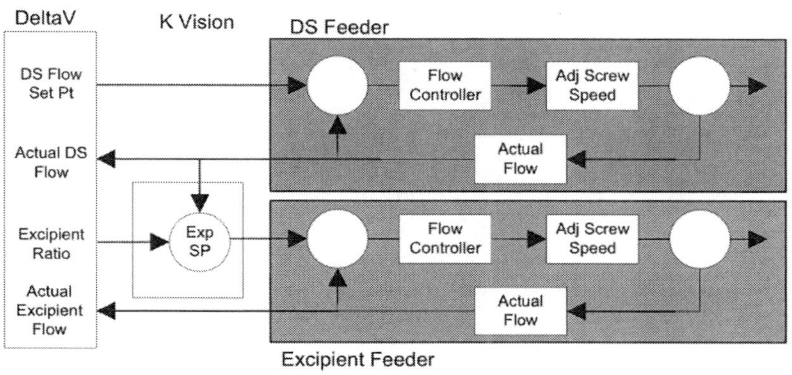

- ALL FEEDERS working together to feed the unit formula reducing short term duration disturbances

11

Ratio Mode Example Using a Poorly Flowing Drug Substance

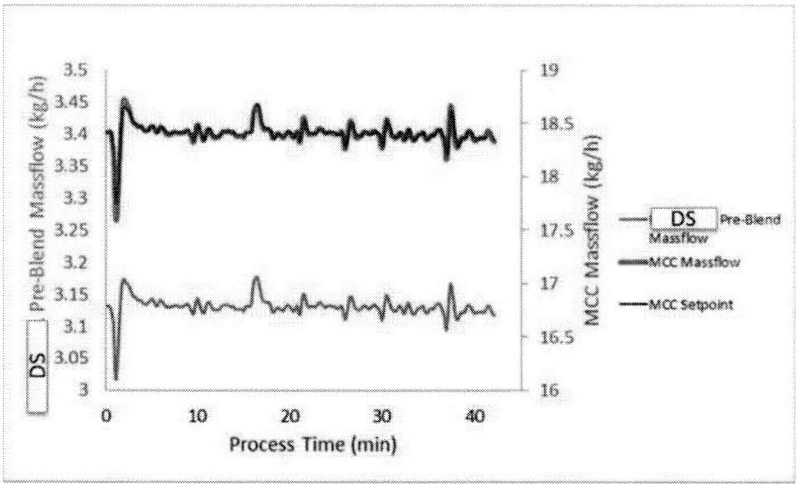

➢ Both common cause (random) variation and deviation from target are judged to be low risks to drug product CQAs when using Ratio Mode.

12

Operating Control Strategy - Feeders

Feedback Process Control

- Individual Feeder Control:
  - Vary screw speed to maintain flow set point
- Feeders Ratio Mode Control:
  - Vary flow set point to maintain unit formula

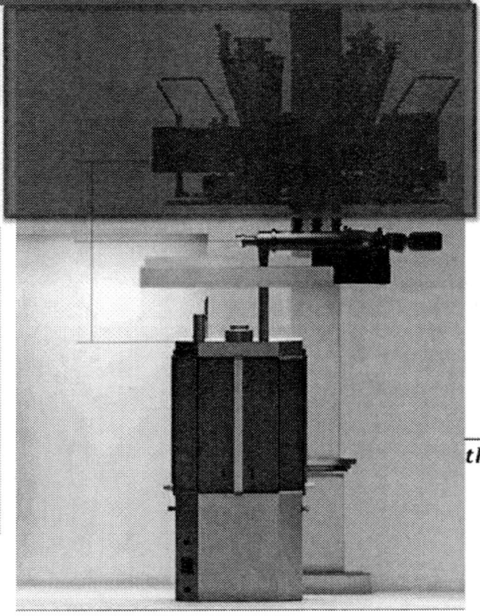

Feed Forward Process Control

Input into the Reduced System Model

Tablet collection/rejection decisions based on fed concentration (% theoretical)

$$\textit{Fed Conc \%Theoretical} = \frac{\frac{\textit{flow rate DS}}{\textit{flow rate total}}}{\textit{theoretical drug load}} * 100$$

13

Operating Control Strategy – Horizontal Mixer

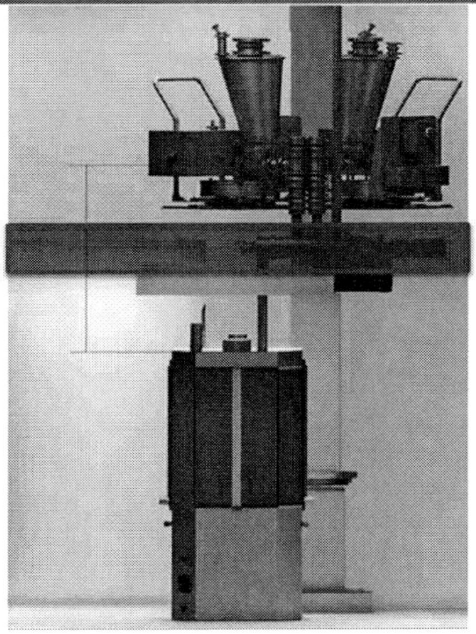

14

What Contributes to Adequacy of Mixing with CDC?

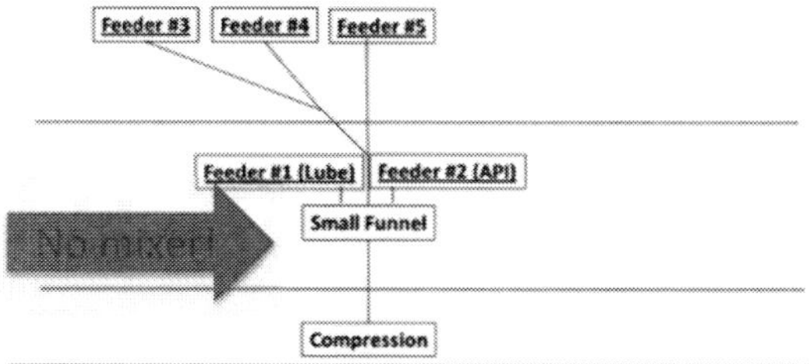

Continuous direct compression manufacturing processes:

- Assurance of uniformity starts with LIW feeders.
- All steps in the integrated process may contribute to mixing.

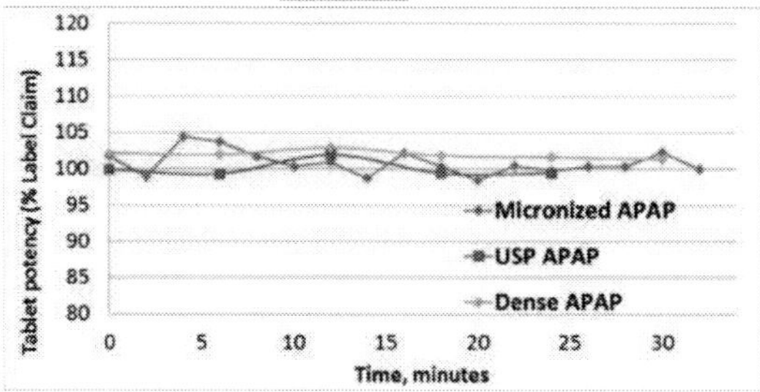

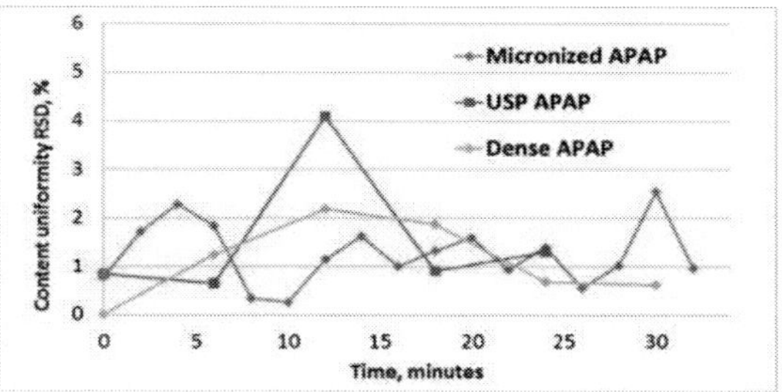

Optimized Design

The appropriate location for assessing adequacy of mixing :

- The outlet of the power-assisted tablet press feed frame,

OR

- The actual compressed tablet cores.

PAT instrumentation:

- A reflectance NIR probe in the tablet press feed frame.

And

- A tablet tester with automated sampling that measures core tablet mass, thickness, breaking force, and transmission NIR spectrum.

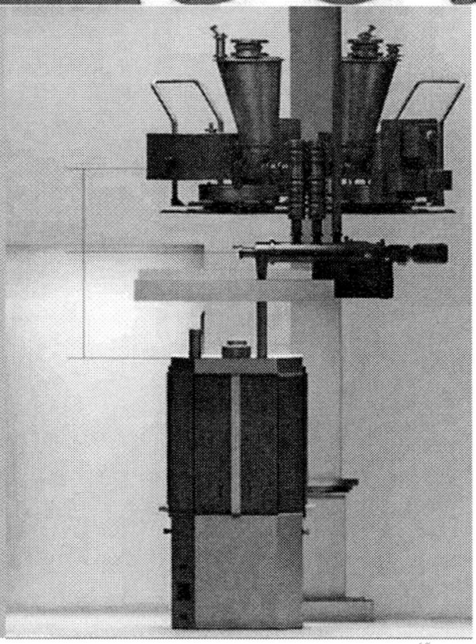

Robustness of Continuous Mixers

- Purpose of horizontal mixer is to dampen LIW feeder variation.
- A mixing DOE was executed to understand horizontal mixing robustness with a challenging formulation (micronized API, relatively low drug load).

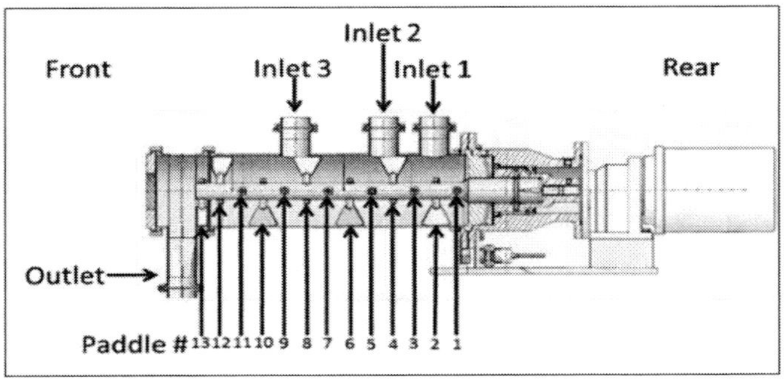

Factors tested:

- Paddle Angle & Direction
- Mixer Speed
- Mass Flow Throughput

17

Mixer and Throughput DOE

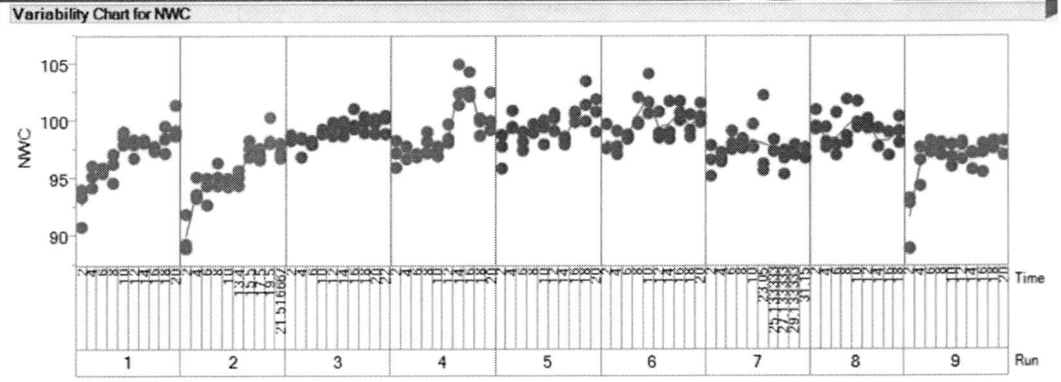

- Mass throughput was the only variable that showed an impact
  - impacted Potency start-up only. No impact to CU, Description or Release Profile.
- Red = Low Throughput; Blue = High Throughput
- Strong correlation of both overall variability and start-up with throughput
  - No other mixing factors were significant with respect to potency and CU

18

Operating Control Strategy – Horizontal Mixer

Robust mixer design and operation.

With appropriate LIW feeder control, horizontal mixer mechanical set up and operation have low risk of impacting drug product CQAs

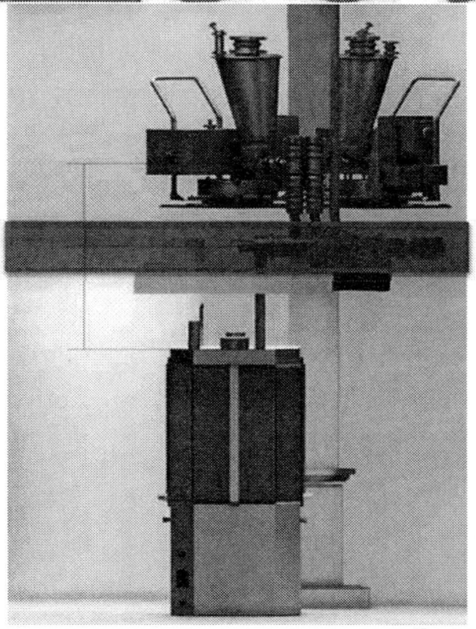

19

Operating Control Strategy – Tablet Press

Leverage the existing internal control loops within the modern, highly sophisticated tablet compression machines (and more!)

20

Operating Control Strategy – Tablet Press

- Tablet Press Force and Weight Control Loops are the same as existing batch processes
- System only accepts tablets within acceptable range of target weight

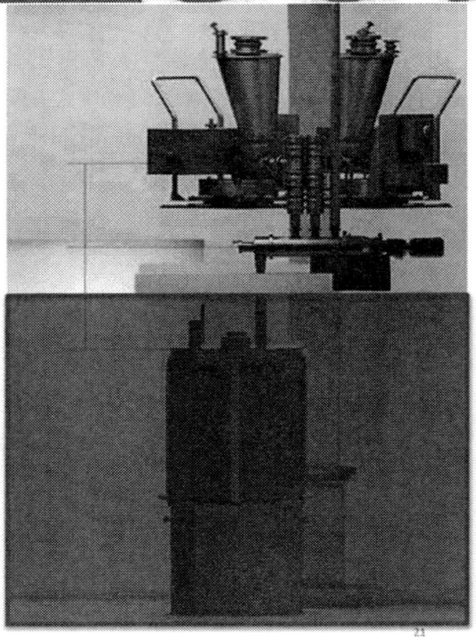

21

Operating Control Strategy – Tablet Press

Data Source	Feedback Process Control
Tablet Press	Tablet weight (Force Control Loop)

Data Source	Feedback Process Control
Tablet Tester	Tablet weight (Weight Control Loop)

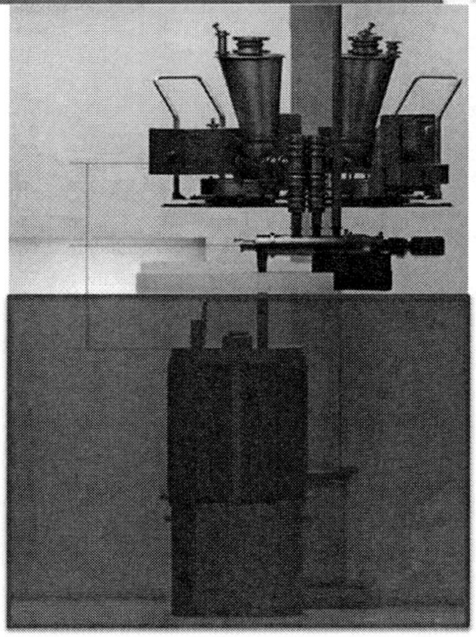

Data Source	Feed Forward Process Control
Tablet Press	Tablet collection/ rejection decisions based on tablet main compression force

Data Source	Option
Tablet Tester (NIR)	RTRt option (UDU/ assay)

Operating Control Strategy – Tablet Press

Data Source	Feedback Process Control
Tablet Press Surge Hopper Level	Tablet press turret speed (Level Control Loop)

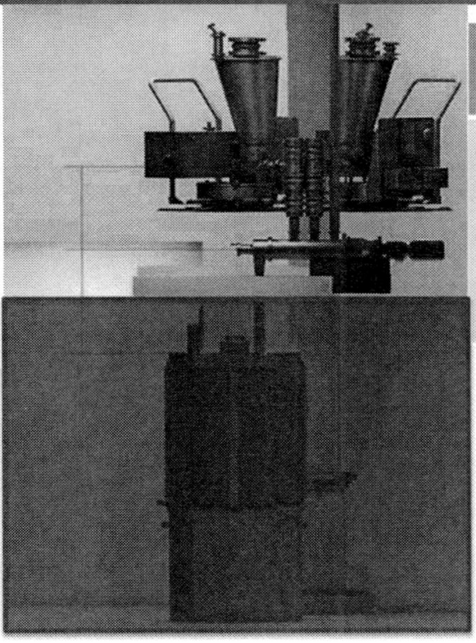

Data Source	Feed Forward Process Control
Tablet Press Surge Hopper Level	Input into Reduced System Model (Time to product and time to clear)

23

Operating Control Strategy – RTD Based Reduced System Model

Leverage the high capability of LIW feeders along with process modeling as part of the operating control strategy.

This is facilitated by the simple physical equipment design.

24

Operating Control Strategy – RTD Based Reduced System Model

Fed Concentration (% ***Theoretical***)
(concentration dispensed into mixer)

RTD based system

Feed Frame Concentration (% ***Theoretical)***
(predicted (future) conc in tablet press feed frame)

Compare to pre-defined collection criteria

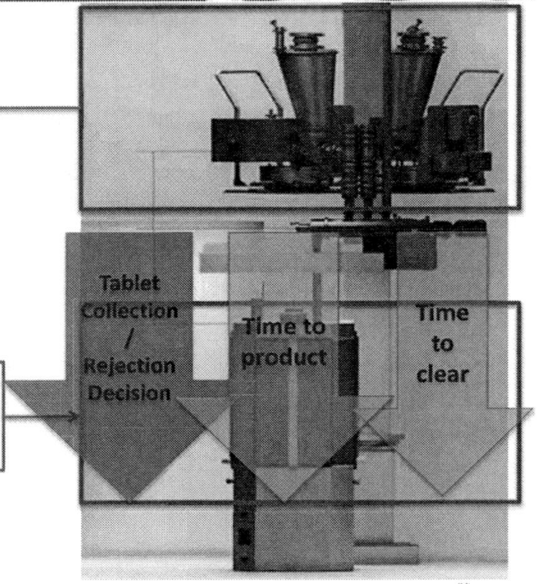

25

Operating Control Strategy – RTD Based Reduced System Model

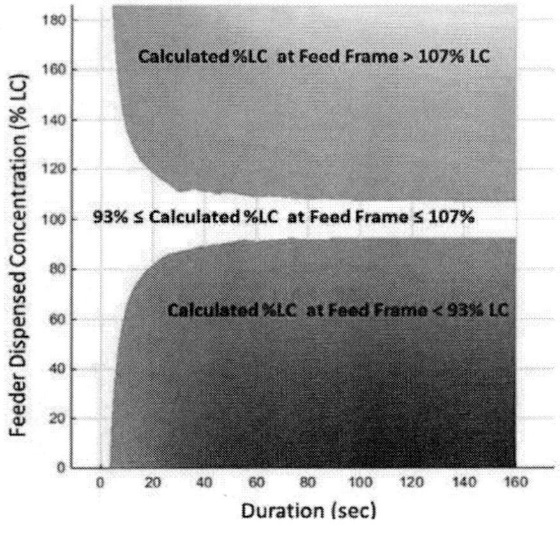

26

Operating Control Strategy – RTD Based Reduced System Model

- Because of the high capability of the LIW feeders, testing the Reduced System Model is similar to automobile safety crash tests; you intentionally trip the system and let it "crash" and see how the system responds to this disturbance.
- Demonstrate process control due to a special cause failure mode.
- For the purpose of this "crash" test, we are using product collection criteria of 93 – 107%

Example – Reduced System Model – LY2, Tablet strength 1

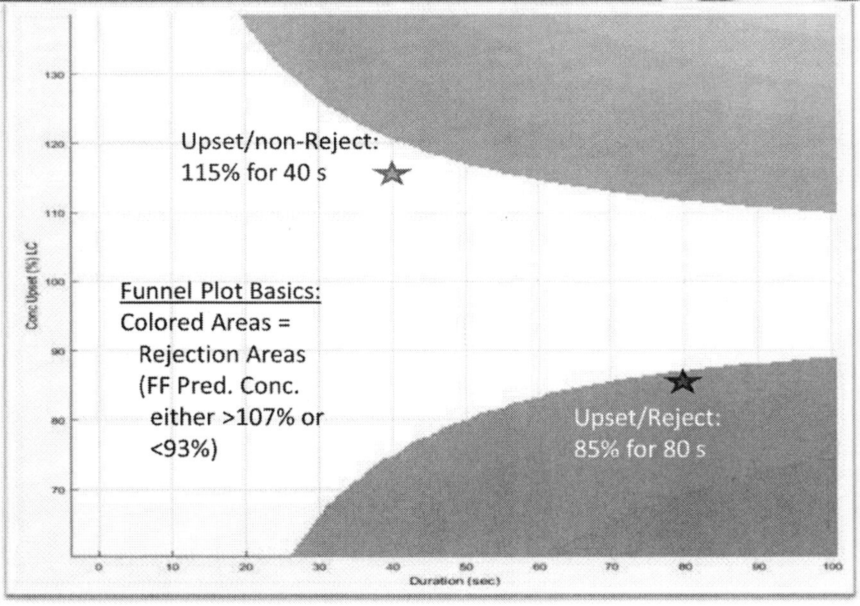

28

Example – Reduced System Model – LY2, Tablet strength 1

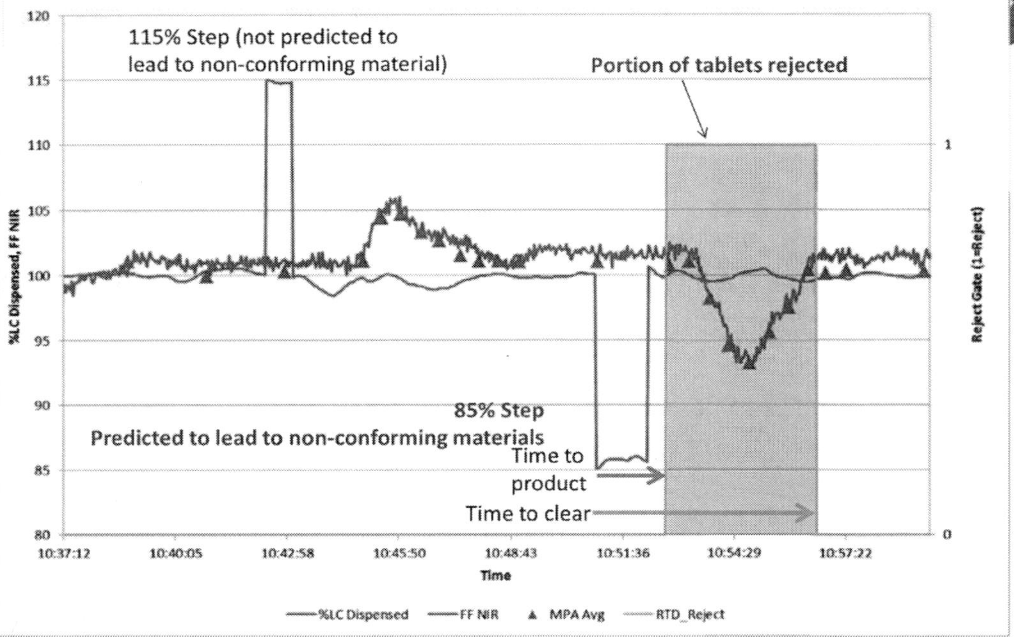

29

Operating Control Strategy – RTD Based Reduced System Model

Data Source	Feed Forward Process Control
Reduced System Model	Tablet collection/rejection based on calculated concentration in tablet press feed frame

30

Are High Frequency Measurement Tools Downstream of the Feeders Needed?

- Based on Lilly's experience in CDC (2013-2016, multiple formulations, sites, campaigns), with our optimized simple design, the majority of special cause/transient disturbances are related to LIW feeders. Managed by:
  - Feeder controls and
  - RTD Based Reduced System Model.
- The need for other tools downstream of feeders are risk based on a project by project basis. The need for those tools downstream of the feeders are related to inconsistent material flow, which was designed out of the optimized equipment configuration.

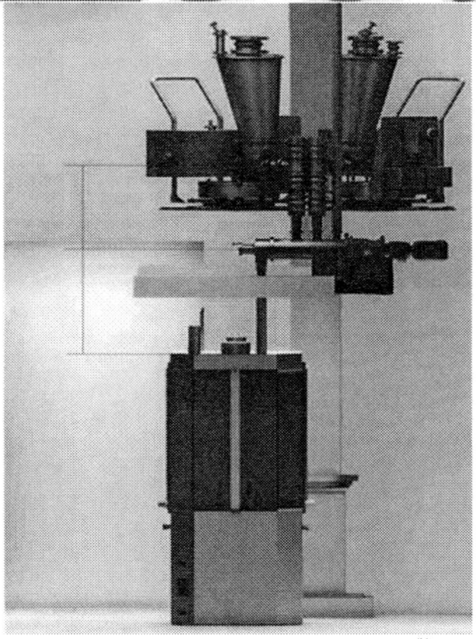

31

Operating Control Strategy – Transient Disturbances

Robust, simple equipment design (minimize occurrence and magnitude of transient disturbances)

Feeder Control

Feedback Process Control: Individual Feeder Control + Feeders Ratio Mode Control

RTD Reduced System Model

Feed Forward Process Control: Tablet collection/rejection based calculated conc in tablet press feed frame

Tablet press control loops

Feedback Process Control	Feed Forward Process Control
Tablet weight Force Control Loop + Weight Control Loop	Tablet collection/rejection based on tablet main compression force

If project specific risks require additional layers (on top of above layers)

Data Source	Feed Forward Process Control Option (if needed)
Feed frame powder NIR (<u>if feasible</u>) and/or Tablet press diagnostics (or other soft sensors)	Process monitoring option If needed, may be used for tablet collection/rejection

32

Recall this example….
Example – Reduced System Model – LY2, Tablet strength 1

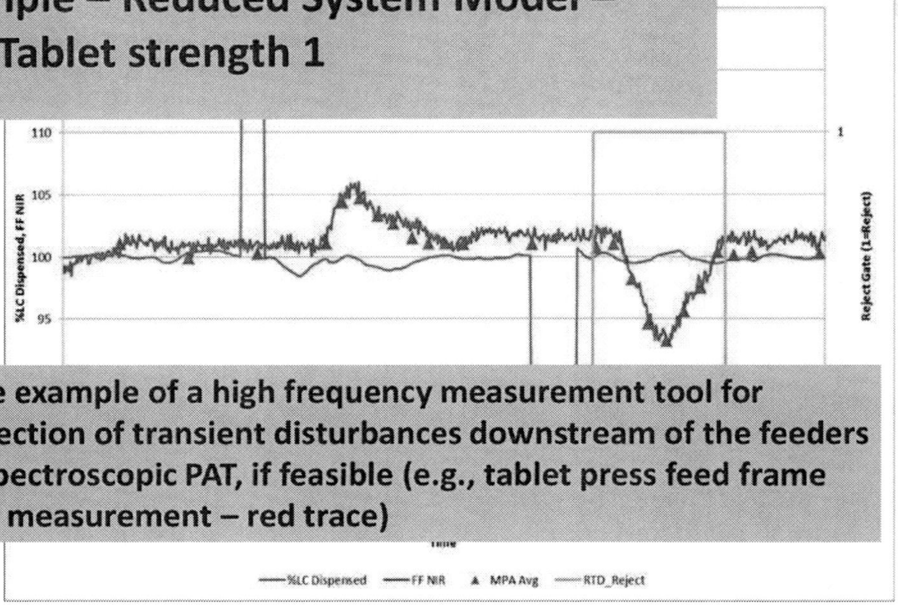

One example of a high frequency measurement tool for detection of transient disturbances downstream of the feeders is spectroscopic PAT, if feasible (e.g., tablet press feed frame NIR measurement – red trace)

33

How About Other High Frequency Measurement Tools?

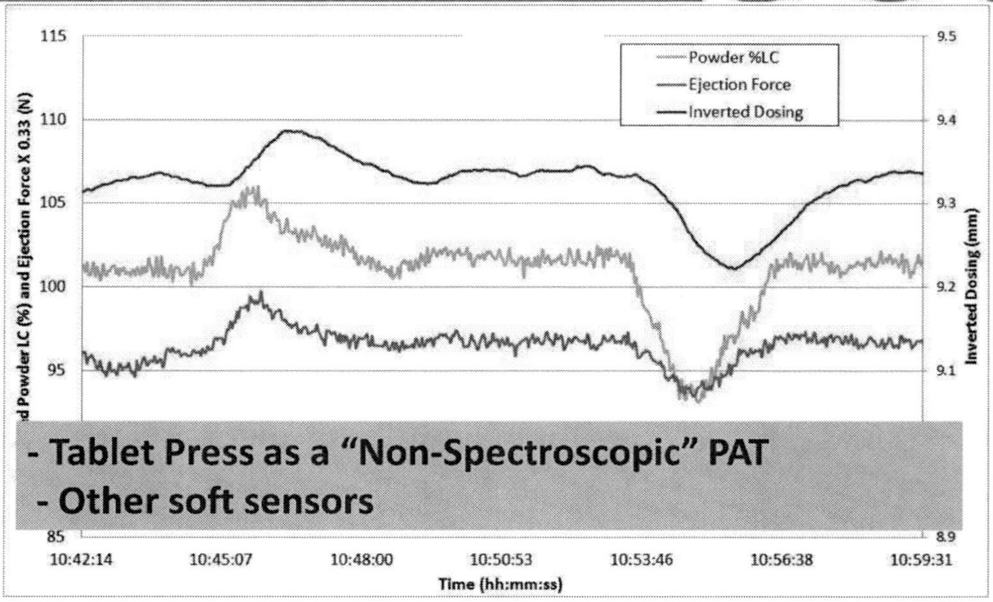

- **Tablet Press as a "Non-Spectroscopic" PAT**
- **Other soft sensors**

34

Improved Process and Quality Control

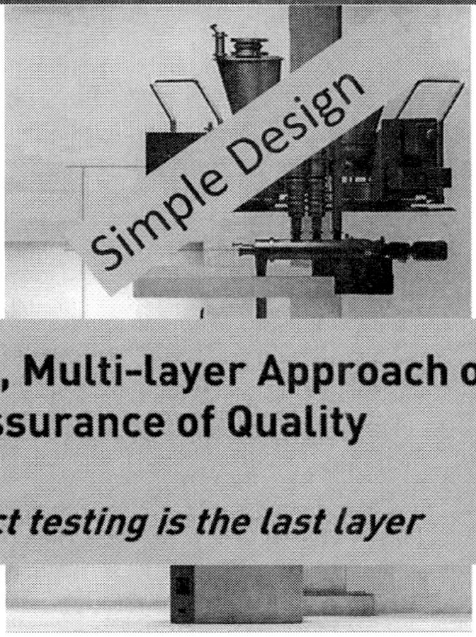

Integrated, Multi-layer Approach of Control – Greater Assurance of Quality

Final product testing is the last layer

35

Acknowledgements

- Ian Leavesley
- Sal Garcia
- Wyatt Roth
- Leo Manley
- Adam Butterbaugh
- Joshua Hanson
- Xiaoyu Zhang
- Alan Mohr
- Lukas Barnes
- Kirk VanDer Kamp
- Jeff Hofer
- Tim Kramer
- Zhenqi Shi
- Bryan Castle
- Evan Hetrick
- Tim Pletcher
- Trent Sommer
- Ron Ruff

Control Strategy, Residence Time Distribution, and Real Time Release Implementation for Continuous Drug Product Manufacturing

Sarah Nielsen, PhD | Senior Scientist, Advanced Technology COE

Jennifer Jacobs, *Stowaway*
Jennifer is a New York based artist
living with Type 1 diabetes.

Acknowledgements

- **Eric Sanchez**
- **Jenny Vargas**
- **Yleana Colon**
- **Elvin Almodovar-Garcia**
- **Efrain Aymat-Liano**
- **Anthony Gonzalez**
- **Gilfredo Navarro**
- **Jeff Kinzer**

Outline

- Process Data
- Control Strategy
- Residence Time Distribution
- In-Line Blend ID for Process Monitoring
- Tablet ID for Real Time Release Testing

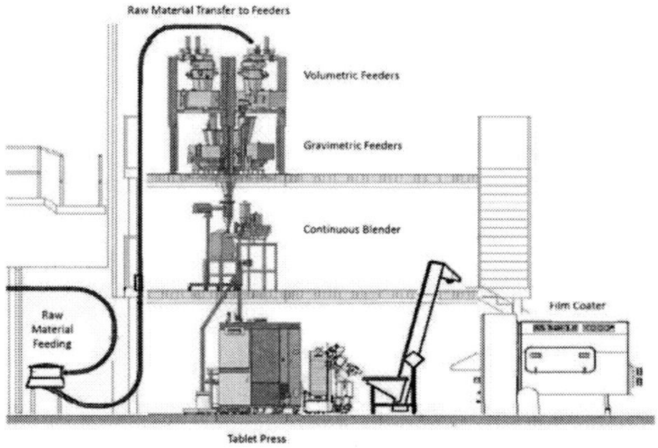

Dry Blend Direct Compression Process

Continuous Manufacturing Process Data Dry Blend Direct Compression Process

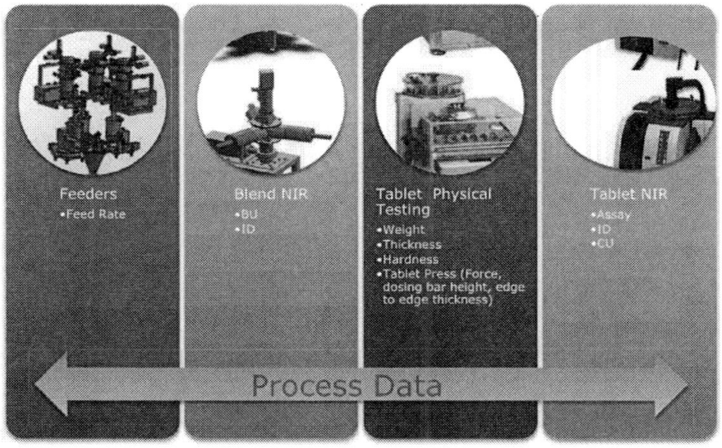

Central control system that defines the methods used to integrate process data

Continuous Manufacturing Process Characterization

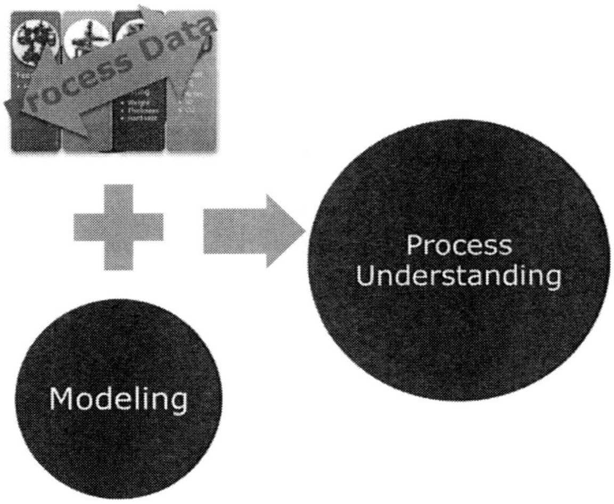

Three Layered Approach

State of Control
- Includes parameters which are monitored frequently to maintain product quality and stability
- Also includes additional redundant in-process checks

Reject
- Disturbance detected with one of the in-process measurements.
- Material is rejected while line comes back to a state of control.

Stop
- Major process problem detected will stop the manufacturing line.

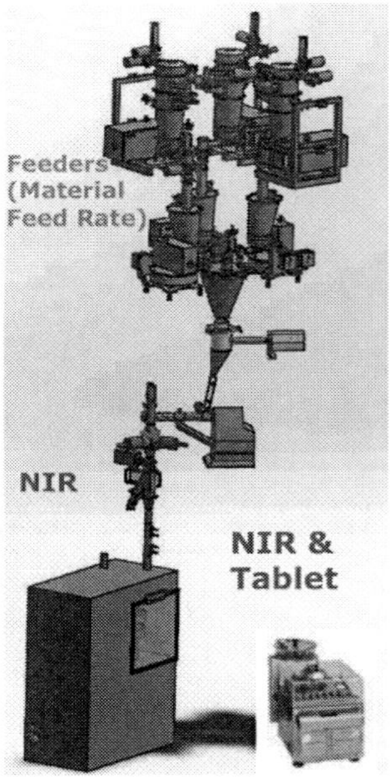

Control Strategy

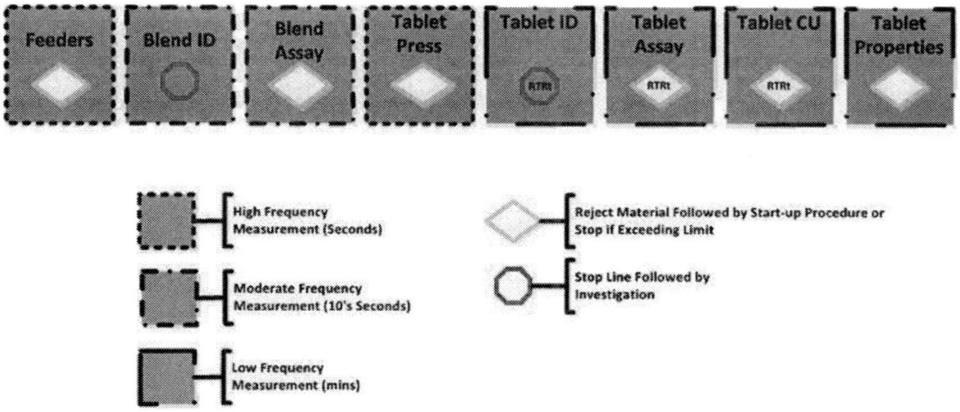

Residence Time Distribution Testing

- RTD is a probability distribution function. Length of time material spends in a system. Mathematically fit to a Taylor dispersion model. Calculate mean residence time (MRT)

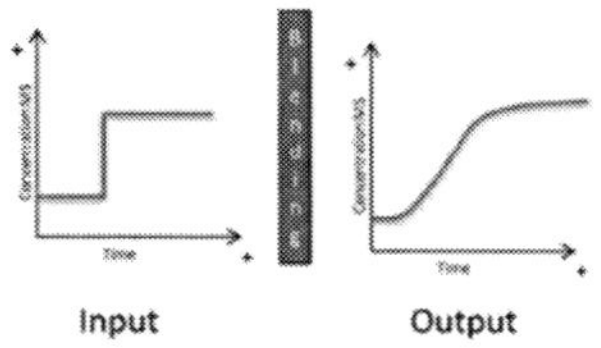

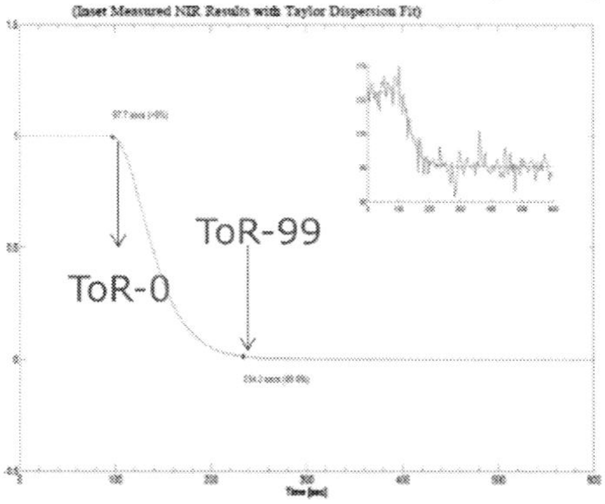

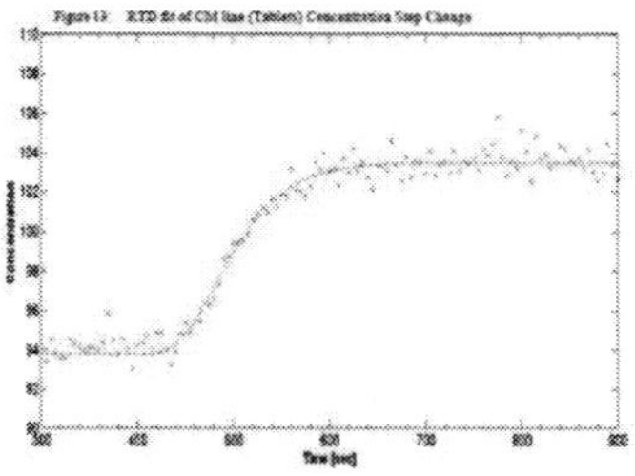

Time from NIR Interface to Diverter Valve 35 Seconds with a 20 second grace period

RTD Timeline of Non-Conforming Rejection

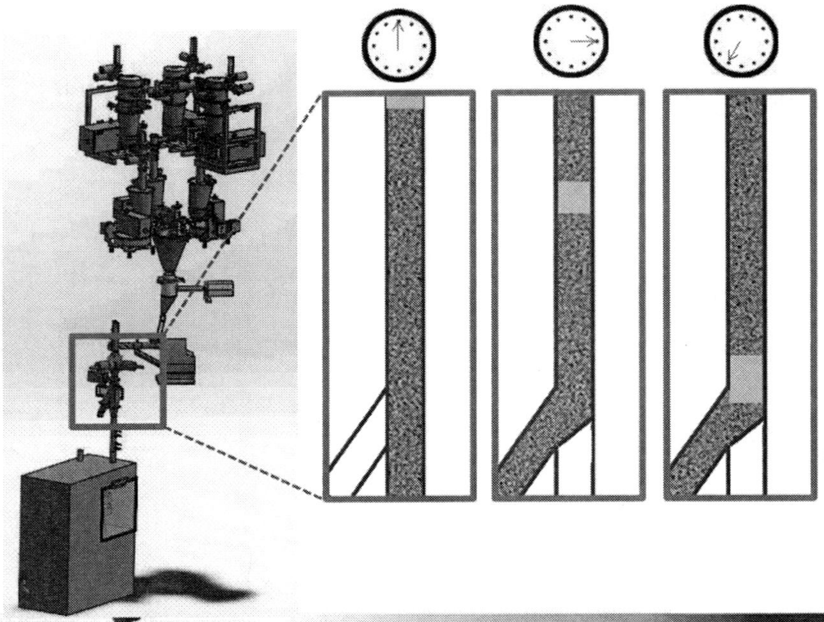

Developing an Automated Blend and Tablet ID Test

- Use NIR spectra and PCA Models X-Space Spectral Residuals to Determine
  - Presence of API
  - Correct API
- X-Space Residual is the lack of model fit statistic

Developing a Real Time Automated ID Model for Blends and Tablets

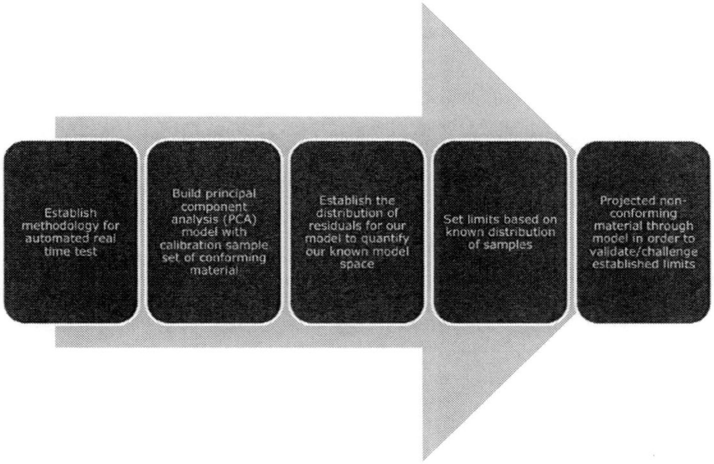

janssen | PHARMACEUTICAL COMPANIES OF Johnson & Johnson

Developing a Real Time Automated ID Model for Blends and Tablets

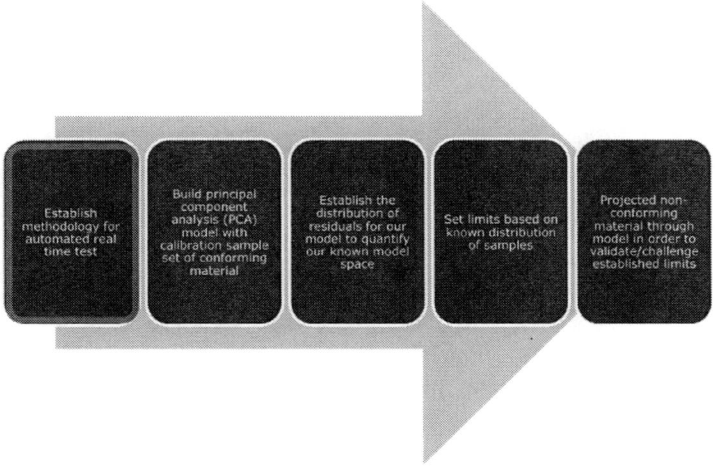

janssen | PHARMACEUTICAL COMPANIES OF Johnson & Johnson

Using PCA to Identify Tablets Outside Defined Model Space

- PCA captures maximum variance in X
- Use PCA to as outlier detection to ID "unusual" sample
- X-space residual statistics are widely used diagnostic tools for out-of-scope sample detection during multivariate model development
- Contributions to X-Space Residuals show how samples are different from PCA model

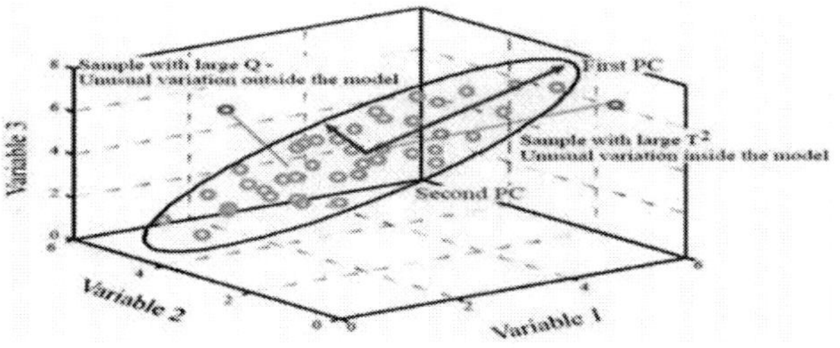

janssen | PHARMACEUTICAL COMPANIES OF Johnson & Johnson

Developing a Real Time Automated ID Model for Blends and Tablets

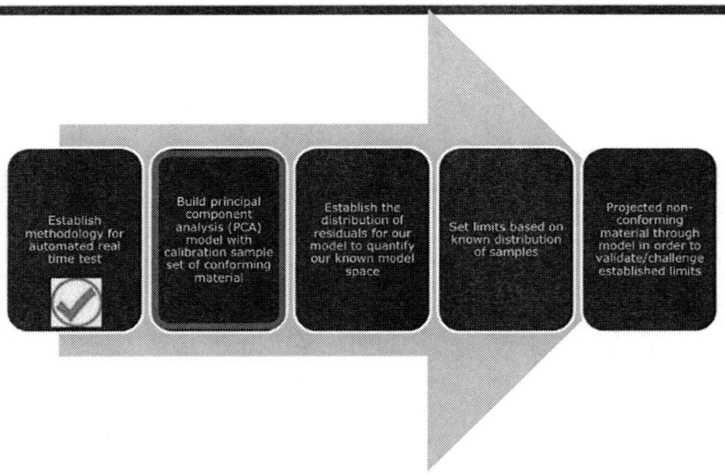

janssen | PHARMACEUTICAL COMPANIES OF Johnson & Johnson

In-Line Chemometric Model Development

Line Run API (%LC)	Line Throughput (kg/hr)
70	40
85	40
100 (API Lot 1)	35,40,45
100 (API Lot 2)	40
115	40
130	40

*Additional samples used to validate model

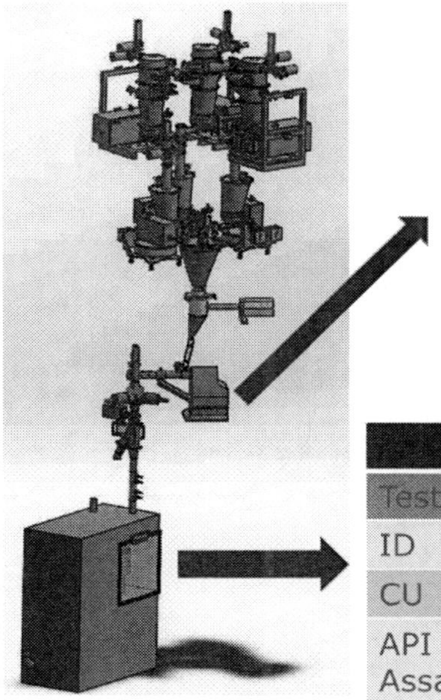

Blend

Test	Model
ID	PCA
BU	PLS

Tablet

Test	Model	RTRt
ID	PCA	Yes
CU	PLS	Yes
API Assay	PLS	Yes

janssen | PHARMACEUTICAL COMPANIES OF Johnson & Johnson

NIR Chemometric PCA Blend Model Development

Parameters	Values
Spectral Region	6032 - 5710 cm^{-1}
Data Pretreatment	SNV + 1st Derivative (25 point window)
Number of scans averaged per spectrum	16
Resolution	16 cm^{-1}
No. of principal components (factors)	2
Concentration Levels in CSS	5
Spectra per Concentration Level (CSS)	10
*Spectra in CSS (API No.1)	50
**Spectra in CSS (API No.2)	10
Total number of spectra in CSS	60

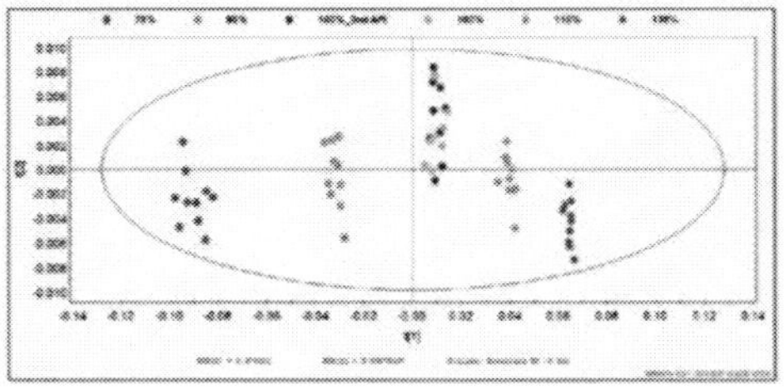

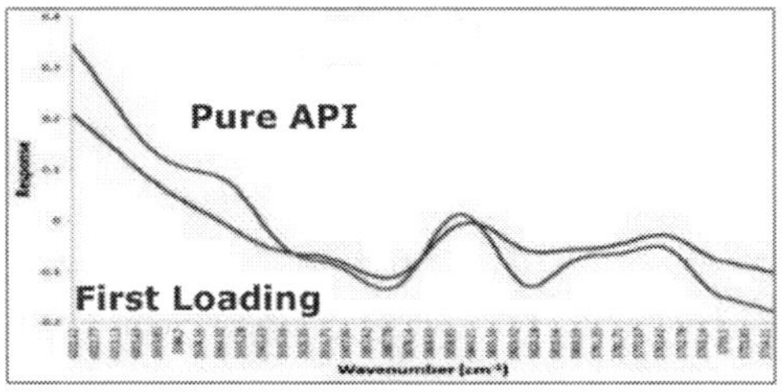

janssen | PHARMACEUTICAL COMPANIES OF Johnson & Johnson

Developing a Real Time Automated ID Model for Blends and Tablets

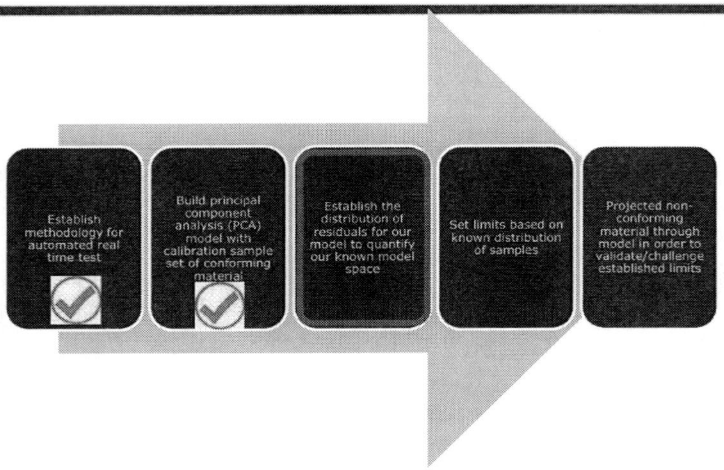

Creating Blend ID Limits

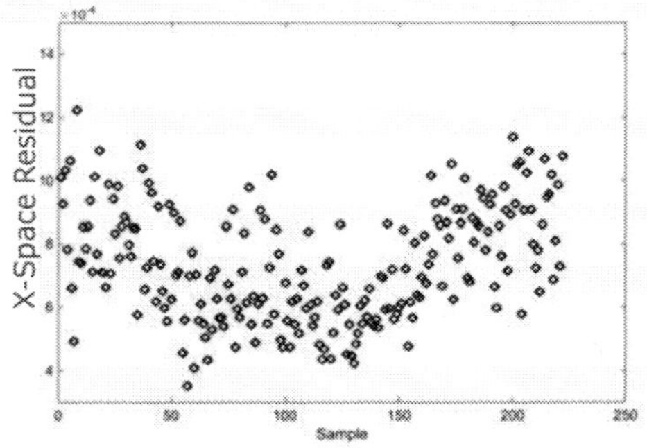

X-Space Residuals of Samples Projected Through Model

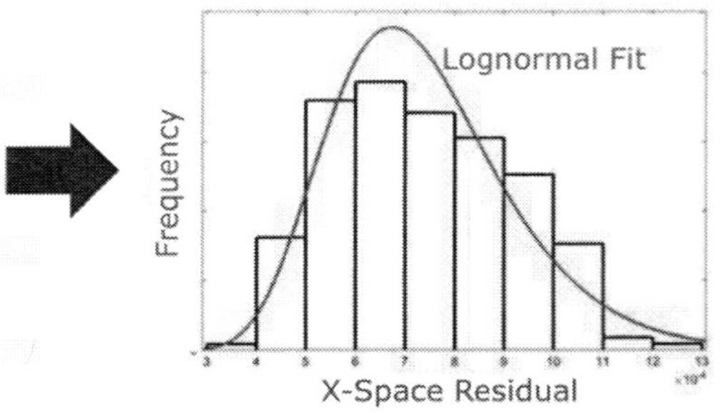

Histogram of Residuals

Rejection Limit Statistics

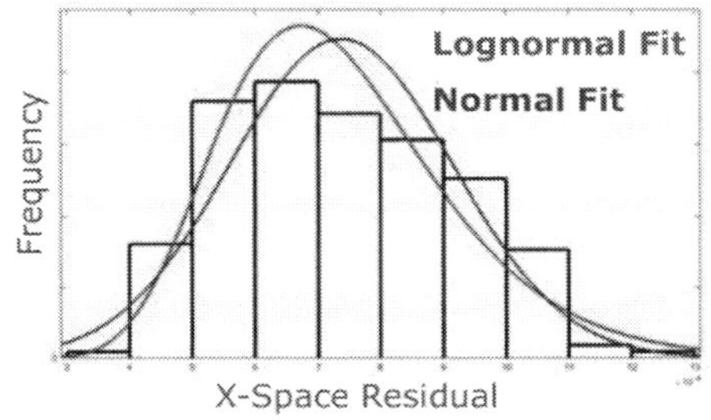

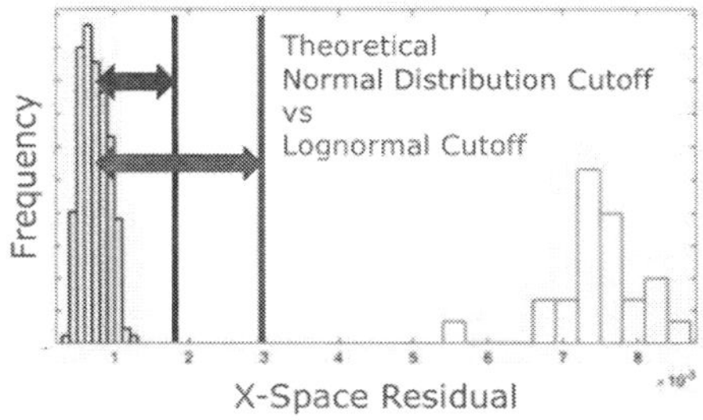

- Appropriate statistics for observed distribution

Developing a Real Time Automated ID Model for Blends and Tablets

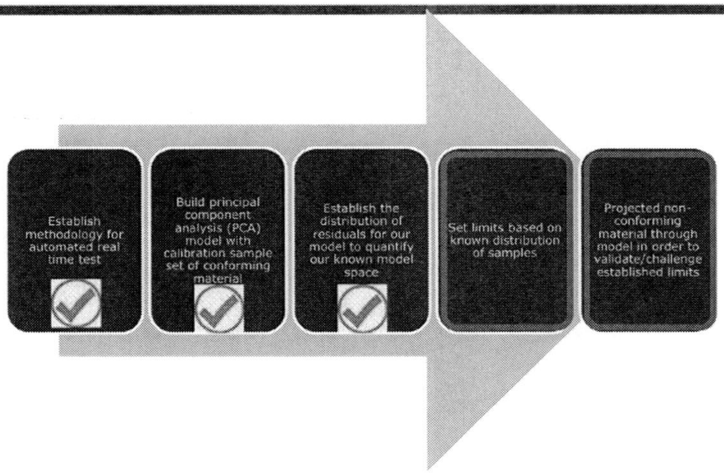

Challenge Samples

- Formulation with correct API lactose swapped for one of the standard excipients
- Placebo Sample
- Correct Formulation with API swapped for a "Foreign API"

janssen | PHARMACEUTICAL COMPANIES OF Johnson & Johnson

Establishing a Limit – Blend Challenge Samples

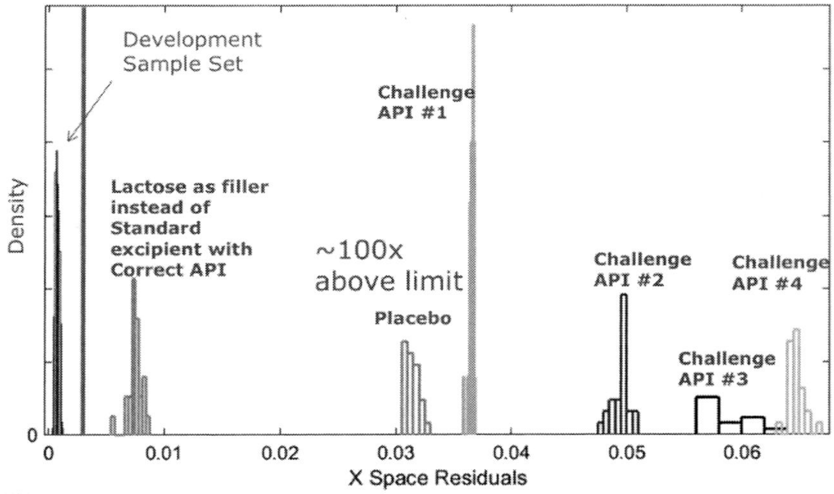

janssen | PHARMACEUTICAL COMPANIES OF Johnson & Johnson

Establishing a Limit for Blend ID

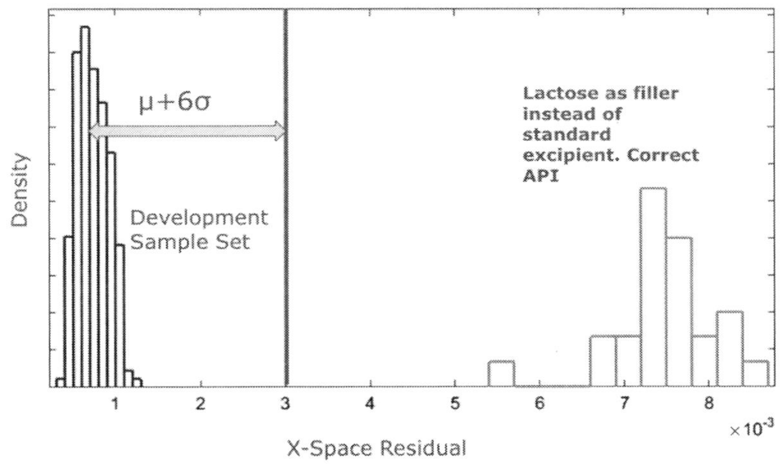

Developing a Real Time Automated ID Model for Blends and Tablets

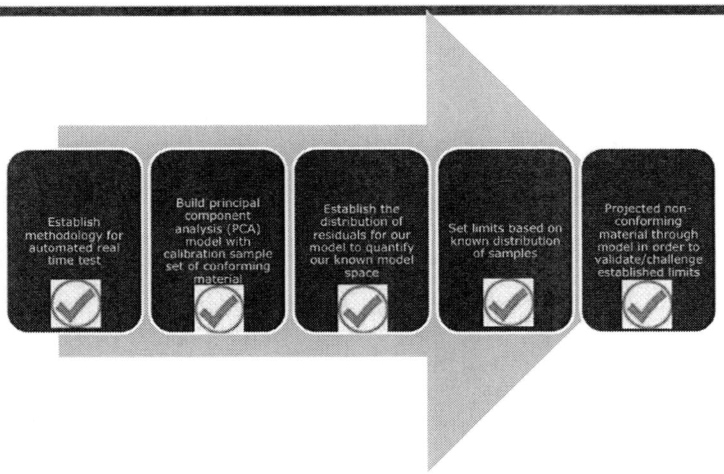

NIR Chemometric PCA Tablet Model Development for RTRt of Tablet ID

Parameters	Values
Spectral Region	12034 – 10900 cm^{-1}
Data Pretreatment	SNV + 1st Derivative (15 point window)
Number of scans averaged per spectrum	32
Resolution	64 cm^{-1}
No. of principal components (factors)	2
Concentration Levels in CSS	5
Tablets per Concentration Level (CSS)	10
Tablets in CSS (API No.1)*	50
Tablets in CSS (API No.2)**	10
Total number of tablets in CSS	60

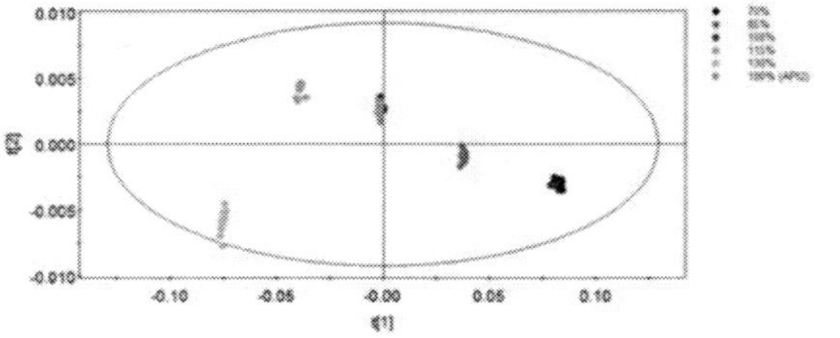

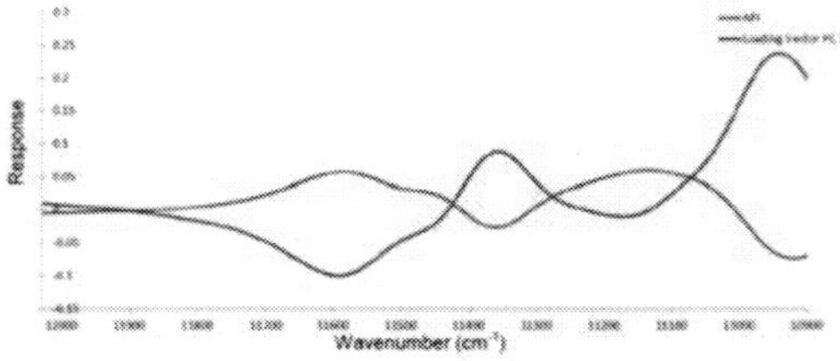

janssen | PHARMACEUTICAL COMPANIES OF Johnson & Johnson

Real Time Release Test for Tablet ID

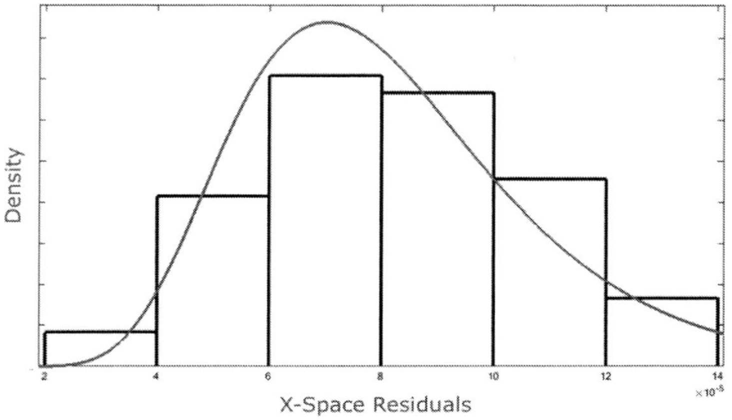

janssen | PHARMACEUTICAL COMPANIES OF Johnson & Johnson

Setting the Limits RTRT Tablet ID

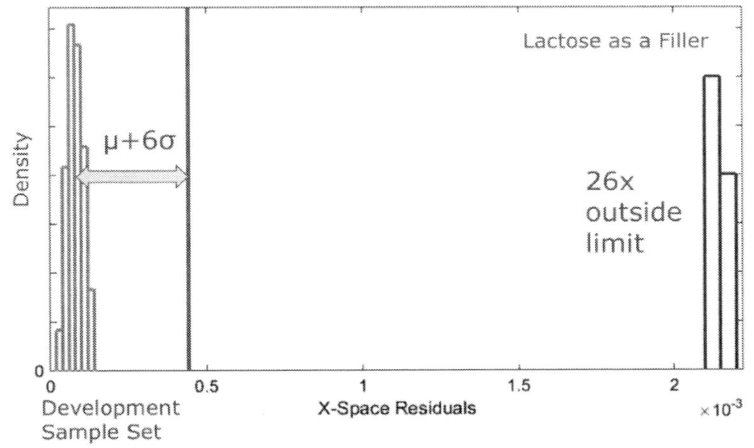

janssen | PHARMACEUTICAL COMPANIES OF Johnson & Johnson

Tablets ID

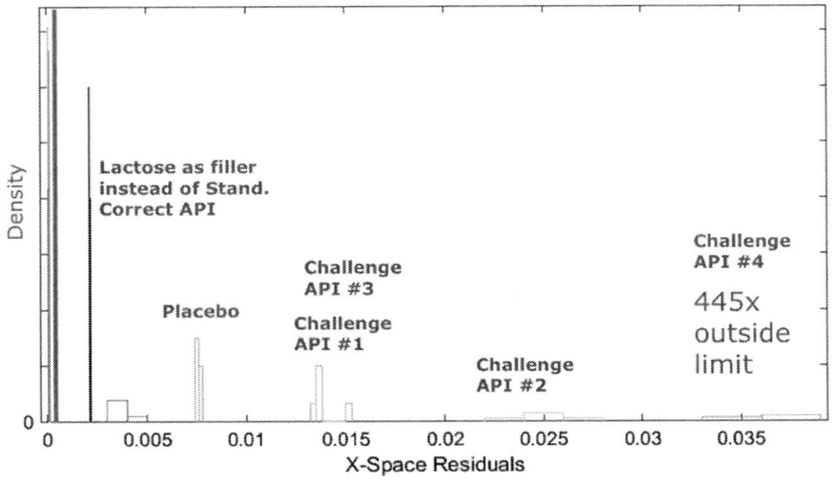

janssen | PHARMACEUTICAL COMPANIES OF Johnson & Johnson

Conclusions

- Established control strategy using PAT as an integral part of monitoring blend and tablet properties
- Accurately determined residence time distribution which enabled real time blend monitoring
- In-Line Blend ID Testing
  - Limits Established using x-space residuals
  - ID Testing confirmed with "challenge blends"
- In-Line Tablet ID Testing
  - Limits Established using x-space residuals
  - ID Testing confirmed with "challenge tablets"
  - Real Time Release Test of Tablet ID

janssen | PHARMACEUTICAL COMPANIES OF Johnson & Johnson

A Decade's Experience Delivering Clinical and Commercial Supplies using Fully Continuous Hot Melt Extrusion

Session V: Continuous Processing and Enhanced Process Control: The Realization of QbD

Robert Meyer, PhD
Colleen Neu
Brandye Smith-Goettler, PhD
Pharmaceutical Commercialization Technology
Merck & Co., Inc.

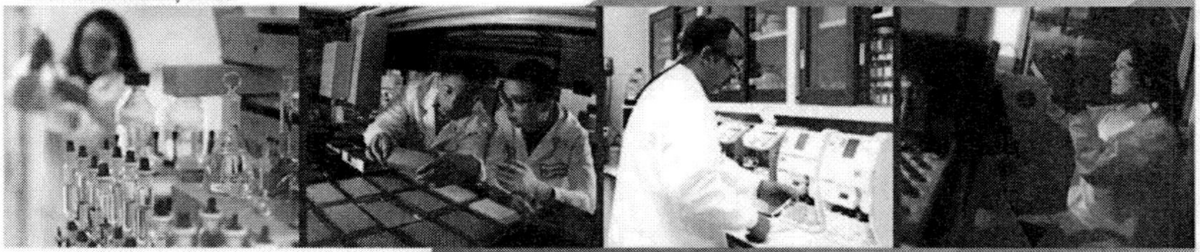

March 2016

Public

MERCK Be well

Presentation Overview

- Brief introduction to hot melt extrusion (HME)
- Merck's HME history
- HME processing options: batch vs. continuous
- Compare/contrast to continuous blending/compression
- Use of PAT for HME in development (RTDs) & production (waste gate)
- Process development for HME
  - Use of development data to establish relationship between process parameters & responses
  - Use of modeling to predict process responses
- Control strategy definition for HME
- Batch definition for HME
- Summary & conclusion
- Acknowledgements

Public

MERCK Be well

Hot Melt Extrusion (HME) Overview

- Hot melt extrusion applications:
  - Generating amorphous solid dispersions
    - solubility enhancement
    - food effect mitigation
  - Controlled release
  - Taste masking
  - Abuse-deterrence

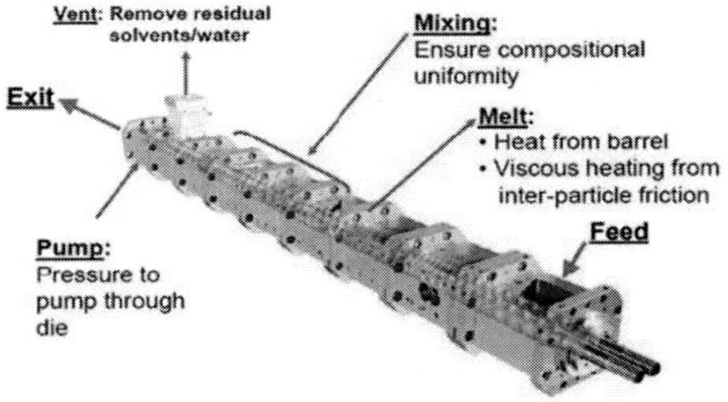

Image courtesy of American Leistritz Extruder Corp

Public

MERCK Be well

Timeline of Solid Dispersions and HME in Industry and Merck

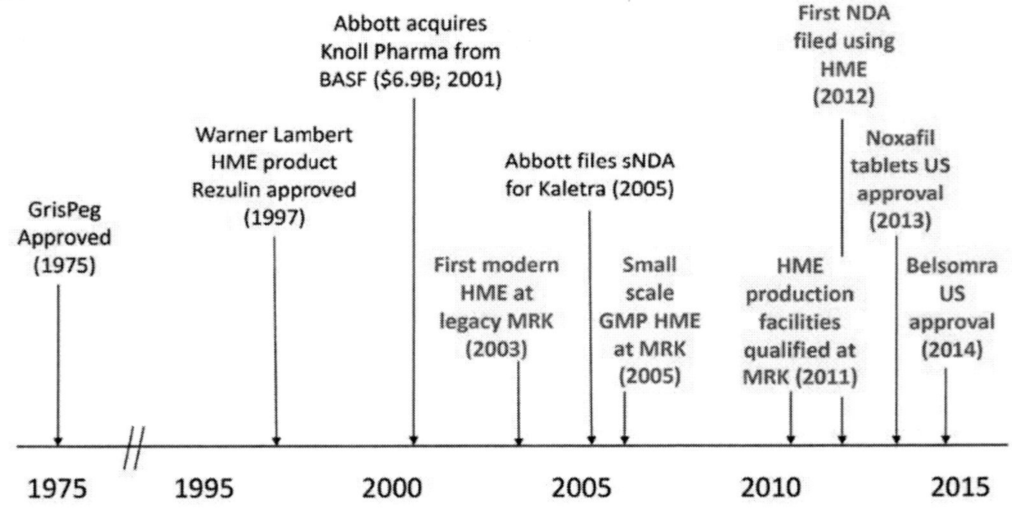

Public

MERCK Be well

Merck HME Products

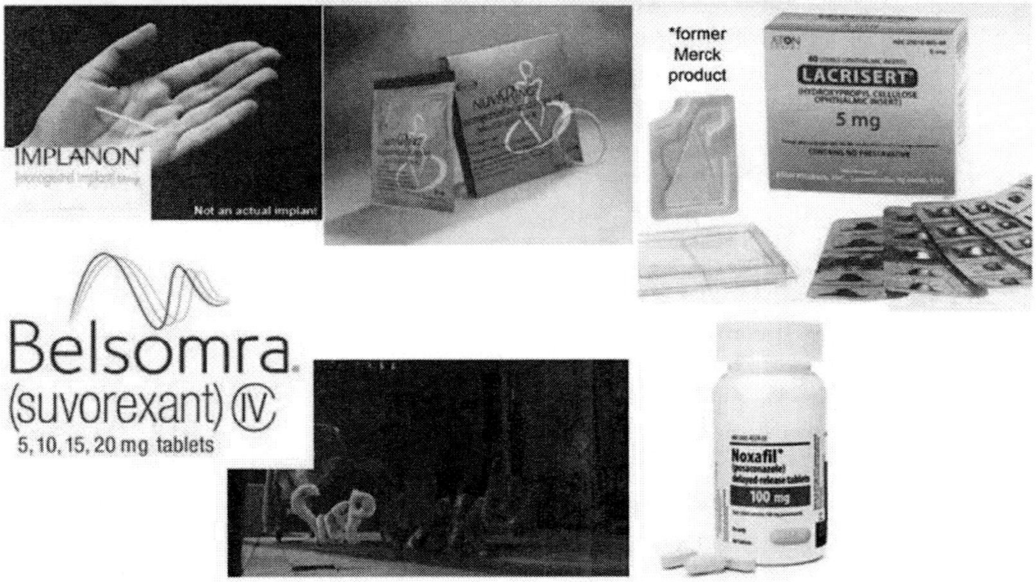

P Public

MERCK Be well

HME Processing Options

- Upstream Process:
  - Continuous feeding of individual components
  - Pre-blending then feeding
- Downstream Process:
  - Milling, Blending, Direct Compression
  - Encapsulation
  - Injection Molding/Direct Shaping
  - Spheronization
- Wide spectrum of options between fully independent unit operations and fully continuous process

P Public

MERCK Be well

HME Upstream Processing: Batch vs. Continuous & Decision to Preblend

- Technical & cost considerations for choosing the upstream process
  - API/polymer feeding trials →
    - Low Tg/highly compressible materials are challenging to feed robustly over long periods of time
    - Preblending = easier feeding
  - Cost analysis of pre-blending versus individual feeding
    - Low-medium volume products were typically cost neutral
    - High volume products benefit from individual feeding

Feeder Performance Degradation
Polymer--- API---

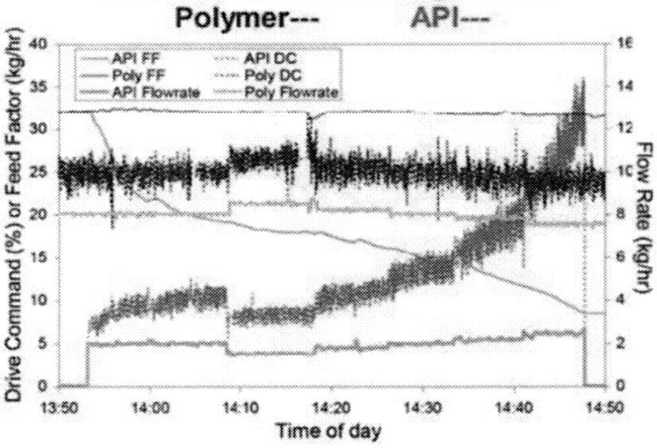

Public

MERCK Be well

HME vs. Continuous Blending/Compression

Similarities to Traditional Continuous Manufacturing	Differences from Traditional Continuous Manufacturing
Residence time distributions leveraged for process development	HME has potential failure modes at each end of the residence time distribution range – inadequate mixing & degradation
PAT utilized for process monitoring/control	Control strategy definition may be different for HME – process responses such as Tmelt used
Batch definition is a key consideration when defining the process	A combination of batch & continuous unit operations may make up the process train

Public

MERCK Be well

PAT Platform for HME – Inline NIR

- NIR signal transmitted through extrudate at exit of extruder
- Online readout of formulation composition every ~5 sec
- Used to detect and isolate bad product during routine production
- Used to measure Residence Time Distributions (RTDs) during process development

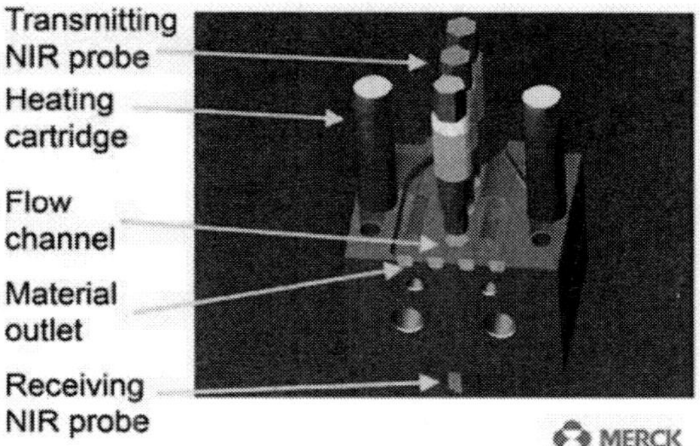

Public

MERCK Be well

Use of PAT During Development – RTD Measurement

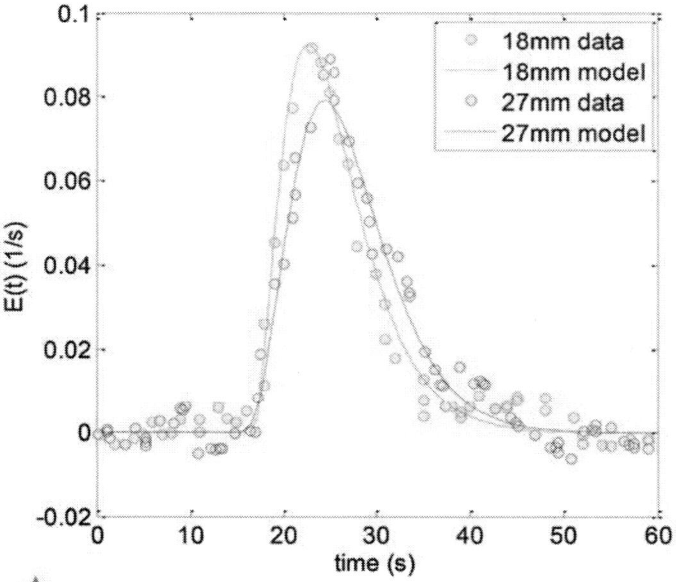

Scale (mm)	18	27
Mass flow rate (kg/hr)	7.0	23.7
Screw speed (rpm)	380	380
Mean residence time (s)	25.7	26.7
StDev residence time (s)	5.4	3.6
Lag time (s)	16.7	15.0
"Mixers in series"	3.1	4.6
Time constant (s)	8.8	12.0

Public

MERCK Be well

Use of PAT During Development – RTD Measurement and Scale Independent Responses

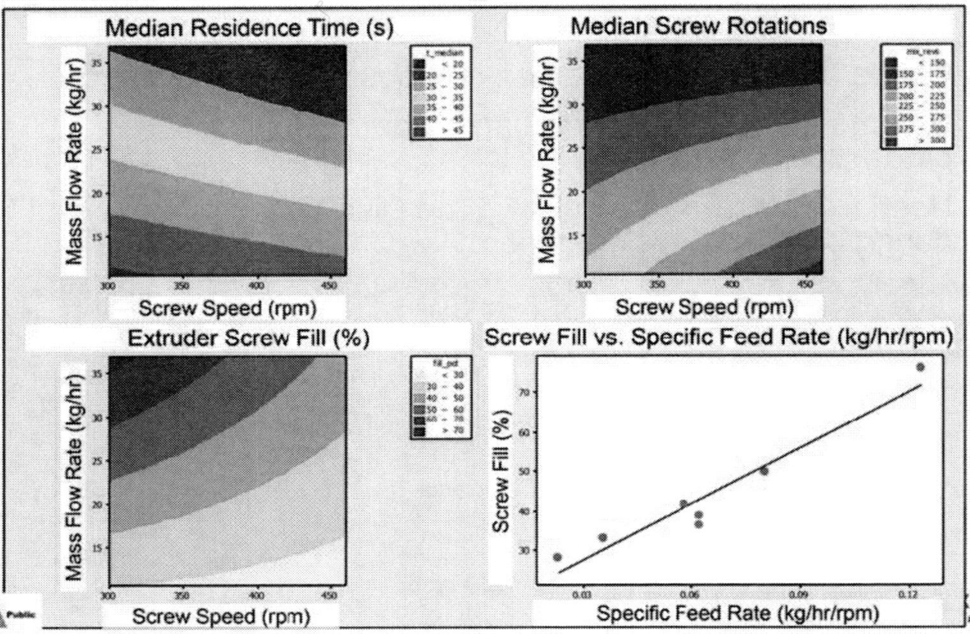

Public

Use of PAT During Production: Waste Gate Control

HME Control Strategy - PAT control of hot melt extrusion waste gate based upon % drug loading (wt/wt)

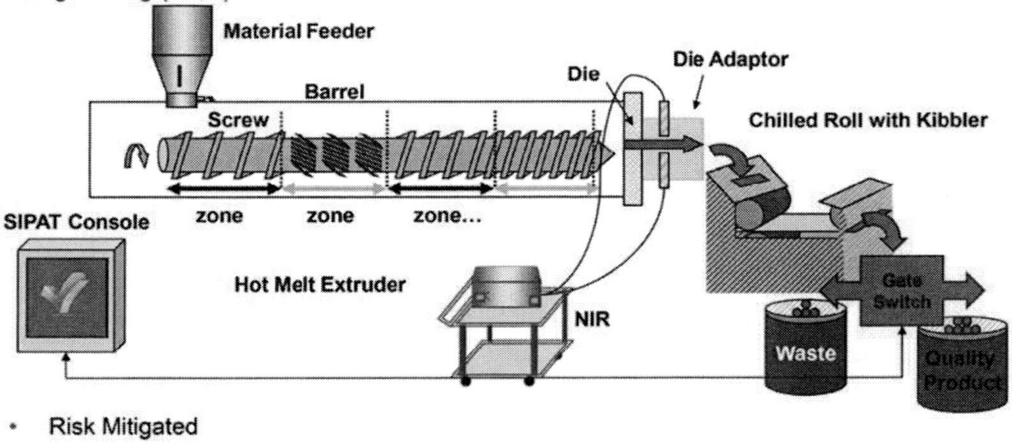

- Risk Mitigated
  - Mitigates risks to composite assay and content uniformity failure within the final coated tablets

Public

MERCK Be well

Use of PAT During Production: Waste Gate Control

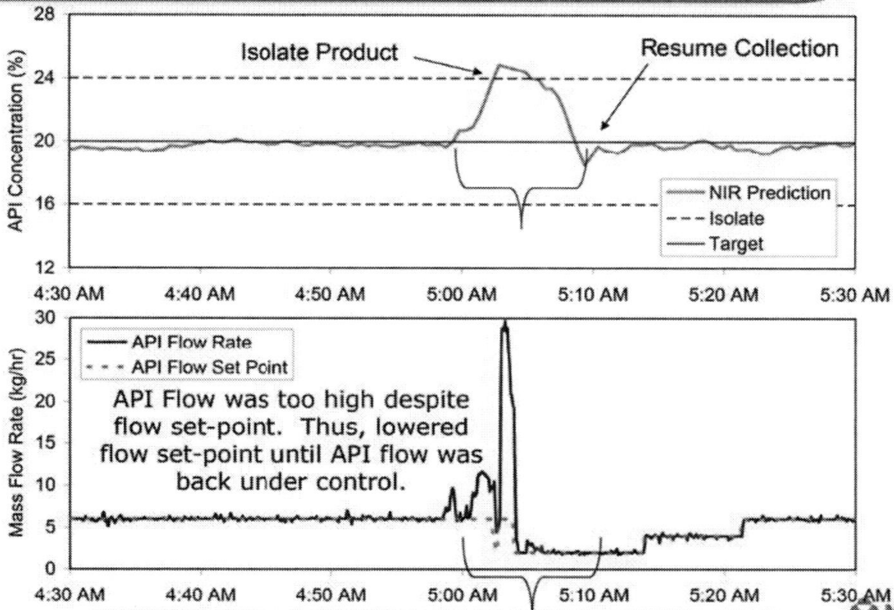

Fault detected and out of spec material diverted to waste

MERCK Be well

PAT for HME – Benefits/Drivers

- **Real time quantitative monitoring and detection of process upsets**
  - Verification of composition/uniformity of extrudate
  - Detect pre-mix uniformity, segregation or preferential sticking issues (or other disturbances – feeder noise, process drift, Pressure fluctuations in liquid feeds)
  - Quality control: isolation of off-specification material
- **"Flying Blind" during mass feeder refills**
  - Mass feeders feed at last known conditions during refills
  - Feeders may need to re-tune for raw material lot changes
- **Enhance understanding of process dynamics**
  - Start up/Shut down
  - Time to reach steady state
- **Provide additional information for residence time distribution (RTD)**
  - Help understand process fingerprint and enable scale-up
  - Design process to get desired quality with fixed feeder performance
- **Holistic/Quality by Design approach for quality assurance**
  - Option to include as part of process control strategy
  - Support continued process verification through monitoring

MERCK Be well

HME Process Development – Linking Quality to Process

Conceptual Representation of Extrusion Process Space:
$\uparrow$ *Throughput &or* $\downarrow$ *Energy = Molecular Disperson Not Achieved*
$\uparrow$ *Energy &or* $\downarrow$ *Throughput = Decomposition*

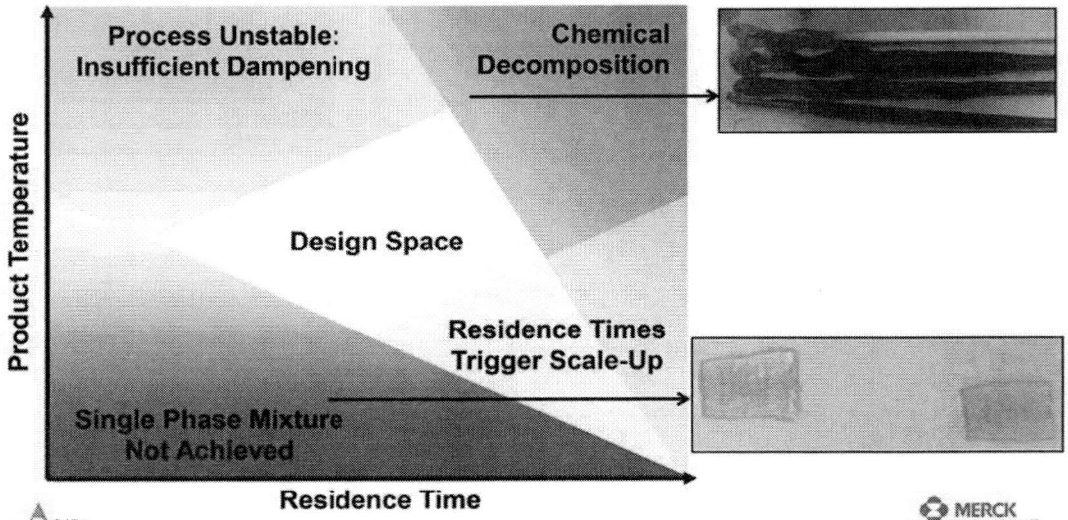

HME Process Development – Linking Quality to Process

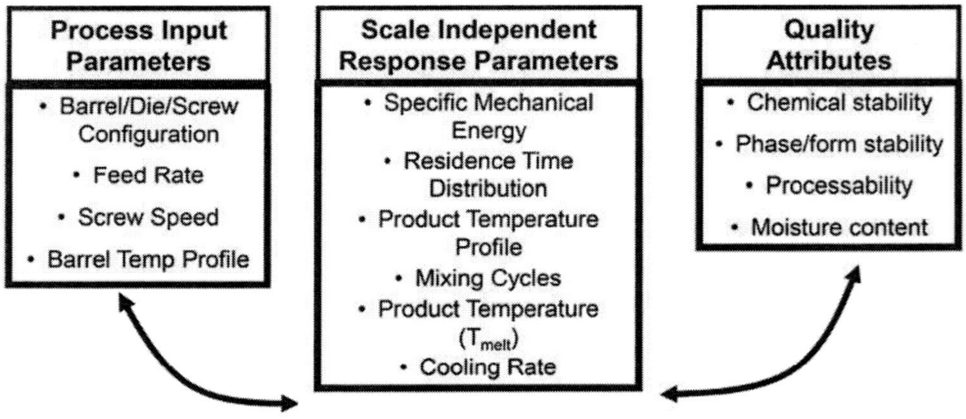

Control Strategy Definition for HME

- Option #1: Use of process parameters
  - Scale dependent, proven acceptable ranges easily defined

Unit Operation	In Process Control	Process Parameter Range	Material Attribute	Test Frequency
Hot melt extrusion	Screw speed	150 – 450 rpm	API concentration (20% ±2% wt/wt)	Every batch
	Barrel temperature	140 – 200 °C		
	Material Throughput	10 – 45 kg/hr		

- Option #2: Use of process response(s)
  - Scale independent, design space needs to be established

Unit Operation	In Process Control	Process Parameter Range	Material Attribute	Test Frequency
Hot melt extrusion	Screw speed	Adjust material throughput, barrel temperature, and screw speed to meet melt temperature and NIRS API concentration measurement	Melt temperature $138 \le T_{melt} \le 213°C$	Every batch
	Barrel temperature		and	
	Material Throughput		API concentration (20% ±2% wt/wt)	Every batch

Public

MERCK Be well

HME Process Development – Relationship Between Parameters & Reponses

- $T_{melt} = f$ (barrel temperature, screw speed, feed rate, screw design, die)

Design-Expert® Software

Tmelt

Design Points

213

138

X1 = B: Specific Feed Rate

X2 = A: Barrel Temp

Product Melt Temperature T_{melt} (°C)

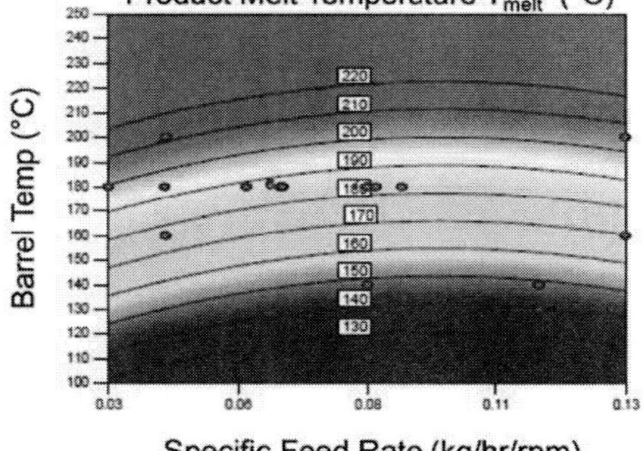

Specific Feed Rate (kg/hr/rpm)

Specific feed rate = (mass flow rate) / (screw speed)

- T_{melt} is only a function of barrel temperature, screw speed, and feed rate once screw design & die type are selected for the process
- Represents design space and translates product attribute back to process parameters

15 Public

MERCK Be well

HME Process Development: Modeling of HME via SCC's Ludovic

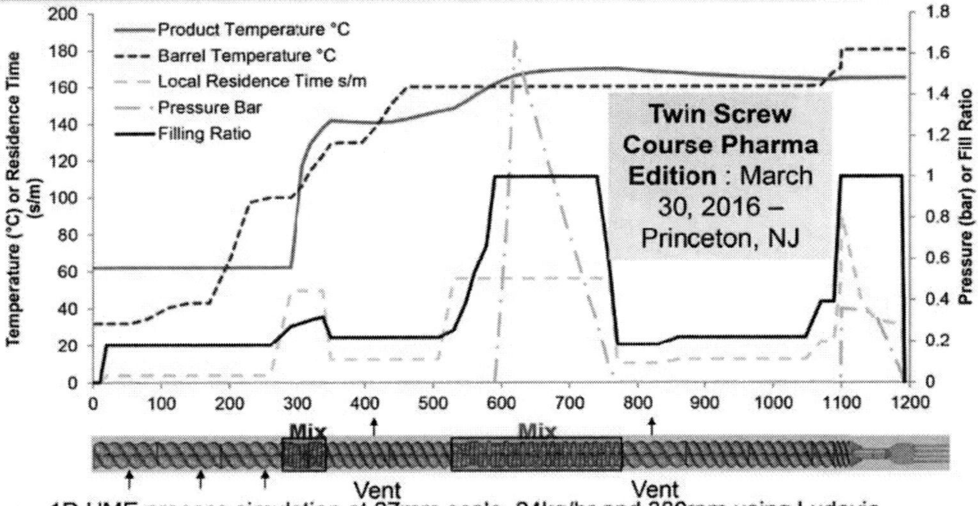

- 1D HME process simulation at 27mm scale, 24kg/hr and 380rpm using Ludovic
  - product temperature changes with screw axial position
  - majority of residence time is spent in the fully filled mixing (kneading block) sections

Modeling moves us from measuring only inlet and outlet to process understanding

Public

MERCK Be well

Batch Size Definition for HME

- Batch size definition could be based on upstream or downstream processing depending on limiting factor(s)
- Individual feed processing will be based on downstream limitations
- Using pre-blends, an extrusion batch can be 1 pre-blend or multiple pre-blends (same process, same control strategy)
- DOM in all cases is the first addition of API to the "batch" – i.e. if 7 pre-blends = 1 HME batch, DOM = API addition to pre-blend #1

Public

MERCK Be well

Batch Size Definition for HME - Example

- Pre-blending = 365 kg in an 1800 L tote (based on pre-blend bulk density)
- HME = 1-X pre-blends - extruded, kibbled, & vacuum transferred to 1800 L totes
- Milling = ≤ 650-675 kg in an 1800 L tote (based on kibbled HME bulk density)

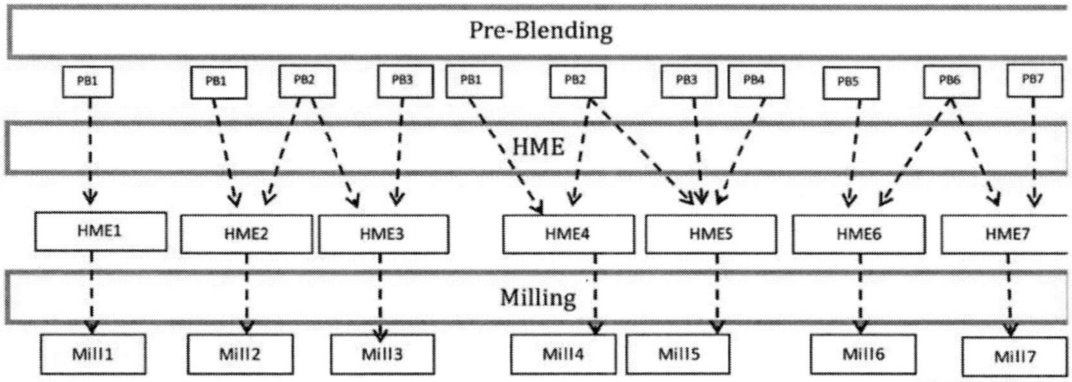

Public

MERCK Be well

Conclusions

- Hot melt extrusion has been utilized for continuous mixing for many years
  - Extent of CP utilized should be balance of technical risks and business benefits
- Enhanced process understanding obtained through QbD development enables attribute based design space
  - First-principles and statistical process modeling contributes to this understanding
- PAT such as in-line transmission NIR leads to stronger process development and process control
- Time interval between new technology lab installation and approved products often takes a decade or more
  - Teams like FDA's Emerging Technology Team and MHRA's Innovation Office have the opportunity to reduce this cycle time

Public

MERCK Be well

Acknowledgements

People

- Alanna Cleary
- Luke Schenck
- Mike Eglesia
- Mike Lowinger
- Colton Bower

Teams

- Hot Melt Extrusion Technology Development Team
- MK-4305 Pharm Working Group
- MK-0859 Pharm Working Group

MERCK
Be well

AUTHOR INDEX